COLLINS GUIDE
TO WILD HABITATS

THLANDS

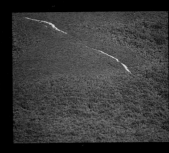

The classic heathland colours

COLLINS GUIDE TO WILD HABITATS

CHRIS PACKHAM

With 60 colour photographs
by the author and 5 colour plates
by Chris Shields

New Forest, a typical heathland

COLLINS
Grafton Street

HEATHLANDS

All this stuff is for the Prettiest Star.

William Collins Sons & Co Ltd
London · Glasgow · Sydney · Auckland
Toronto · Johannesburg
Text and photographs © 1989 Chris Packham
Colour plates © 1989 Chris Shields

First Edition 1989

Designer: Caroline Hill

ISBN 0 00 219844 4 Paperback
ISBN 0 00 219866 5 Hardback

Filmset by Ace Filmsetting Ltd, Frome,
Somerset
Colour origination by
Wace Litho, Birmingham, UK

Printed and bound by
New Interlitho SpA, Milan, Italy

CONTENTS

RIGHT **Weeting Heath, Norfolk**

Acknowledgements are due to: John Buckley, for putting science in my head; Nigel Webb and Peter Merrett of Furzebrook Research Station and the Stour ringing group for access to their knowledge; to Claire Buckley and Dave Scott for ruminating and regurgitating my drafts; to Andy Welch and Barbara Levy for compassionate tolerance beyond the call of duty.

And especially to: Stephen Bolwell for never-ending hedonistic encouragement; to my parents for allowing me to pollute their freezers and conversations for years; to Joanne for sleeping in my car all over England; and to Jenny for the rails.

ACKNOWLEDGEMENTS

LEFT **A heathland pool in summer**

1. PICTURES OF THE WRECKAGE

The mauve half-light that precedes dawn cloaked my vision, and the softest of dews chilled my knees as I tumbled through a fence into a piece of history – a piece of land that our time has almost forgotten. I lay there, flat on my belly, stealing back some of my neolithic heritage for my senses, staring across the steppe watching pieces of flint hop randomly like the rabbits they turned out to be. It was flat and broken, like a pockmarked billiard table covered with biscuit crumbs. Fescue and flints ruined by rabbits and ragworts. It appears in my memory that it was silent, but in reality I must have heard the distant bark and whirr of a Pheasant and the background trill of a Skylark. The monochromy and lack of perspectives peeled away as the light flared up over my shoulder and washed away the trace of mists over the heath.

Smell, sight and hearing blossomed as the dream developed into reality. The smell was dusty, musty, sandy and dry. The sight was a relict of the grassy heaths, a summer dawn in Suffolk, England. The sound was the scream that had lured me many miles – the most evocative, compelling, priceless sound that I can strain out of the chaos and wreckage of our ruined wildlife. It's sharp, distinctive, eerie; its fiendish feel sends shivers of neat romantic nostalgia down the spines of even long dead naturalists. Eruptive bursts of peeling, wailing, ringing whistles lead to a polyphonic cacophony as one pair of these high priests of the steppe encourages an excited wailing with other pairs of Stone Curlews. Whisper, whisper, whistle . . . Kur – LEE, Kur – LEE, Kur – LEE. . . .

By seven o'clock on this June morning the aura of heathland had gone bittersweet on all my senses. The Stone Curlews' shrillings were replaced by the deafening songs of Whitethroats, I was blinded by the brilliant burning sunspots on a stream and

LEFT **Heathland's past, a ram's skull**
RIGHT **The skeletons of burnt gorse**

9

Red-backed Shrike, the prettiest pastel bandit

something inside me was biting, burning and kicking. I could feel this little piece of England exploding all through my head.

I crept down the stream's bank, soaked by the soaking sedges overhanging the path, pausing only to watch an origami frog spasm across the jade water surface. He drifted away, guided by the stream's geography, not by his flicking feet. Pipits spun up and over the bracken and a ruddy Stonechat perched on top of the blazing gorse, whose coconut perfume was beginning to rise as I pushed my way through. It would later be heady and sickly, but now the day was still brittle sharp.

The clearing in front of me was filled with a clump of Sloe bushes buried in small brambles and gorse, the early sun shattered their tops and I winced through a confetti of down and insects to survey this bowl. This was an oasis of rarity fit for a prince. I scanned every twig, every bush top and bramble spray around the glade. A

rabbit slipped away and a Brimstone butterfly burned like a sulphur firework in the shade of the sloes. And then I had him. The elixir. The jewel.

My world condensed around this one avian hub. Nothing stirred, was even alive outside my heartbeat and the male Red-backed Shrike. A little pastel bandit. A living jewel on the crown of oblivious, over-exposed blossom. He was perfection, beauty and pretty pricelessness. He was one of the last. His little tail flicked, he raised his breast, he twitched his dainty head.

I couldn't see him clearly. I mean not every feather, nor even any feather. He was patches of colour, whose shapes changed with his reactions and attitudes. Yet the subtlety of his grey, black and russet hues, of his structure, shape and line was for me the mastery of adaptive design. He was loafing, warming up like some reptile. Nothing more really happened; he may

Dunwich Heath, Suffolk; a palette of heathland colours

have flicked his face with his feet, I can't remember; he didn't fly until the magic had gone. I've never seen this shrike again, although I've searched this little heath every year since. The oasis still sparkles, and the energy of heathland still hums in tune with the insects; but the Red-backed Shrike is almost extinct in England. The jewel has gone. He's left behind him the spangled sand lizards, the twinkling and tumbling Silver-studded Blues, the rasping Dartford Warblers, the rascalous Adders and the heat, the dust and the wreckage of heathland.

HEATHLAND DEVELOPMENT AND DESTRUCTION

The heathland habitat as we know it only occurs in north-west Europe, although other forms of heath occur in North America, Canada, Russia and Scandinavia, and extensively in South Africa and Australia. Heathland's appearance is dependent on

suitable conditions of the soils and climate providing a scope for the development of a 'dwarf shrub community'. The European heaths have ericaceous shrubs, primarily Ling *Calluna vulgaris*, and principally includes the lands which border the North Sea, English Channel and Atlantic coasts. Strewn from the north-west coast of Spain, along the west coast of France, through Brittany and Normandy to Belgium, through the Netherlands and across the north German plain, up into Denmark and terminally in southern Norway and Sweden we can find the relics of our heathlands. The whole of the British Isles falls in this area. At altitudes above 300 m heathland becomes moorland; in this book we are concerned only with the lowland heaths and their distinctive flora and fauna. Despite the oceanic, cool temperate climate and poor acidic sandy soils it is generally accepted that some form of woodland is the climax community in the above biogeographical zone, so where did all the heathland come from?

Once upon a time, 4,000 years ago, the sun cracked through the endless forests of Elms and Limes and woodpeckers laughed as some neolithic man hacked at a huge tree with his sad flint axe. But by the time his burial mound was full there was a clearing in the forest, and his great-grandson's cattle scuffed in the rich, brown soil of the clearing as the sun burned down. The next time the vegetation in that clearing was hacked at, the axe was made of bronze, the trees fell quicker, and by summer a stand of tatty oats stood full of Corn Cockles and Corn Marigolds. By the time the Celts arrived, raping and pillaging their way over Europe, the clearing was becoming barren, grassy and dusty; it had been eroded and rabbits had begun to breed. When the Romans arrived their carts bumped over the now huge rabbit warrens and sheep scattered across the plain. Augustus wiped the sand from his eyes and could see for miles. The Domesday Book recorded shifting cultivation, and no forest at all. There were no woodpeckers left to laugh now, but Great Bustards turned themselves inside-out to display, and in the south Dartford Warblers and shrikes shouted at the furze cutters.

This was heathland, and it was man-made (with a little help from the rabbits and sheep). It later became common land and peasants fought for the right to hive bees, to graze animals, and to cut fuel. But the fences went up, the manors formed, and sheep and rabbits in their millions chewed on any tree that dared to grow on the heath. Later when a peasant dug his way out of his sandy hovel in the nineteenth century he noticed that there were fewer sheep, and when he got back from Waterloo his neighbours and sons were at work in the towns. The peat stacks were green with moss, and there was a black rock burning on the hearth. So now the furze on the heath grew tall and in autumn yellowing birches glowed in the sunsets. By the early twentieth century the forest was growing

Damp down on the heath

Sunset over the New Forest, Hampshire

again, the plough cutting and the fires burning. Our civilization had made heathland and now it was going to destroy it.

The ploughs worked day and night during the wars, and the sands of Suffolk blew under the bellies of Flying Fortresses. Nearly all the rabbits died of myxomatosis and the heath grew taller still. In the south, cities sprawled and cut the wastelands into a million pieces. So now, as I stand overlooking Lakenheath with my father, and survey that slaughterfield below with the last of the Red-backed Shrikes, we've come from Spitfires to space shuttles, bakelite to silicon, from Glen Miller to the Sex Pistols in less than fifty years. There are no sheep and as I pick up a worked flint outside an old rabbit hole, an F-111 roars overhead, and I wonder what will be left in fifty years' time for the survivors of the AIDS generation.

Depressing. It does seem that heathlands may have suffered even more of our mistreatment than other habitats;

but it may be that they were simply less numerous to start with, or that they were more loved, written about and explored by naturalists past and present, and therefore now more missed. Anyway, despite this decay, it must be said that some of the species who make their homes here do show a remarkably stoic tenacity to survive in spite of us. They exist in the shadow of oblivion on the wreckage of the wastelands. Just such an example are those crepuscular lunatics the Stone Curlews.

THE STONE CURLEW
Everything about these guys is weird. Sound, movement, eyes, behaviour. They are like living fossils damned to survive out of their time. They are the birds of the Danish invasions, of Stonehenge, of Crécy, crazed reincarnations of pagan virgins, the Knights Templar and Boudicca. Stone Curlews move like tin soldiers in a cartoon with a faulty freeze-frame facility. Running fast, with short spurts and then

13

stops. Complete stops. For seconds. Frozen like terrestrial herons. Then a short, slow, deliberate walk. Then a stab, as they pluck a beetle or a grasshopper off the turf. At other times they feed more like plovers, wandering about regularly picking at insects for prolonged periods. In fact, analysis of their parasitology, biochemistry and skeleton, and the down-pattern of their young leads taxonomists to class their closest relatives as the plovers *Charadriidae*, rather than the Bustards *Otitidae* which they more superficially resemble and with whom they once shared their Breckland homelands. Now our remaining British Stone Curlews are confined to the arable deserts on the Hampshire and Wiltshire downs, and the sandy Suffolk heaths.

Stone Curlews are nocturnal and, being especially active at dawn and dusk, their huge, vacuous eyes are conspicuously adapted to such an existence. A piercing chrome yellow, they have a very contractile iris, which opens as daylight fades to allow much improved vision. They are gregarious birds and even during the breeding season groups meet up in the twilight for their bizarre displays and communal feeding activities. Their vocal and visual repertoire is entertaining and confusing. Upright, stiff-legged strutting with fanned tails is a sign that birds are in conflict: the antagonists pace alongside each other and run about frenetically – they may even engage in bird-to-bird tussling. A curious jousting occurs when angered pairs run at each other and peck out angrily as they pass. They then run on a little and turn, repeating the ritual two or three times.

Stone Curlews' courtship antics are even less understood. It is thought that northern European breeders may be already coupled by the time they return from their Iberian winter breaks. They have been seen in groups, running, jumping and skipping with ruffled feathers and flailing wings, necks twisting and arching. This ungainly arching forward of the head is more often associated with the selection of the nest site and subsequent scrape making. It is also a greeting ceremony when one bird returns to the other on the nest to take over the incubation, the returning bird often presenting the other with a small object as part of the process. The nest is merely an enlarged scrape lined with small stones and rabbit droppings, made by both sexes in April on the open ground. Towards the end of the month, the hen lays two cryptic eggs, each with her own individual frequency and density of markings; producing a pattern that seems consistent from year to year.

These fascinating avians have been on the decrease for years. Their numbers must have peaked during the 1700s, when the sheep-walks and commons were in their halcyon days and the brecks, downs and plains of Europe were largely complete. In the last fifty years the decline has been measurable, and is largely attributed to relentless habitat loss. Old breeding haunts have been afforested or thrown over to arable cultivation, and now a third of the eggs laid are destroyed by farm machinery. Pesticides too must have played a prominent part, since the Stone Curlews' food has been shown to comprise earthworms, beetles, spiders and woodlice, all of which would have held high levels of toxins. Carrion and larger prey are also taken. One female was seen feeding half-grown young on the intestines of a myxomatosed rabbit, killing a Lapwing chick, and eating a clutch of Skylark eggs. The birds seem to fly off mysteriously to feed elsewhere under cover of darkness. If this is generally the case, then a couple of postage stamps of breeding ground are clearly inadequate for the species to thrive: suitable satellite feeding sites are also needed. For the moment, though, they return to the heaths at dawn to display, before lapsing into diurnal lethargy to pant in the heat haze of the baking sands, squinting sleepily through their big eyes.

RIGHT **A pair of Stone Curlews change over nesting duties at sunset**

By definition heathland is open and uncultivated, a bare, and more or less flat tract of land, covered in a low herbage of dwarf shrubs, especially Ling *Caliuna vulgaris* and *Erica* species of heather, which overlay poor soils. The soils are acid, nutrient-starved, and almost universally this Ling, which is sometimes called heather, is the dominant plant, often occurring in a monotonous monoculture. Evergreen, and adapted to a variety of climatic conditions, it's a low and woody species which grew initially in forest clearings; as the forests were felled and the soil exposed in the moist oceanic climate, with its tolerance of these rotten soils, Ling has flourished. It has a high reproductive ability, large plants producing hundreds of thousands of minute, highly fertile seeds which are dispersed by the wind and can last for years in the soil waiting to germinate when conditions are favourable. Also, since heathland is such a new habitat, man-made and maintained since the Neolithic times, there has simply not been enough time for many species to evolve to exploit it. Thus it is a very species-poor community – one bird, which needs gorse as well (Dartford Warbler), seven or eight moths (Emperor, Fox etc.) and a handful of other insect herbivores restricted to the dominating heathers. Combine this with heather's slow growth and strength as a floral competitor, and the result is a very boring overall flora and fauna in terms of diversity.

In Britain, and indeed on continental heaths, this problem is at least aesthetically resolved by the presence of species from further south in Europe. They occur here on the northern edge of their ranges because the heat, the dust, and the heathland habitat structure are somehow comparable to their mosaic of southerly habitat types such as scrubs, hedgerows, woodland edges or dusty Mediterranean

LEFT **A thick blanket of Water's Crowsfoot**
RIGHT **Wild Gladioli**

Spiked Speedwell, a breckland rarity

plains. Sand Lizards, Dartford Warblers, Stone Curlews, Tiger Beetles, Smooth Snakes and the Ladybird Spider are all restricted to heathland in Britain, and they are rarities because of this. More southerly continental plants can also be found, although many are now seen only at restricted sites in the Brecklands of eastern England.

These brecks are, or were, the grassy heaths; large, flat and sandy areas where rabbits and sheep heavily grazed the sward of fescue and *Agrostis* grasses in a semi-continental climate. This resulted in a steppe-like terrain, and the over-grazing and erosion led to huge inland dune systems which once buried villages in swirling sandstorms. There is only one such dune left today, Wangford Warren near Lakenheath airbase. You would be hard pushed to bury Trafalgar Square in this paltry patch, but it does illustrate the basics of the habitat, the mossy and lichenous carpeting and graphical sand sedging surviving like a textbook history you can tread on and touch.

A PROCESSION OF PLANTS

One of the continental colonists, the Spanish Catchfly *Silene otites*, can be found near here. It is a stickily hairy perennial which catches insects in the same way as the rest of the other European catchflies, but looks nothing like them. It is unbranched and has a rosette of dark green leaves with delicate spikes of yellow-green flowers; it occasionally reaches 60 cm in height. Its relative, the Sand Catchfly *Silene conica*, is a tiny, neat, downy greystemmed plant which shares its taste in open disturbed soils. When I found one of these pretty little pink-flowered plants I was amazed at its pathetic roots. As a sand dweller I presumed it would make some attempt to anchor itself. It had failed entirely and lay flat on the sand exposing a small tap-like root about 1 cm long.

The Spiked Speedwell *Veronica spicata* once flowered freely on the open grass heath, but now it cannot compete in the deeper sward that has taken over in many places. Its long flower-spikes are packed with intense blue flowers in July, but along with the Spring and Breckland Speedwells *V. verna* and *V. praecox*, and a host of other rarities, it is restricted to a few secret sites and very difficult to see.

It was the aftermath of a fire that once provided me with a fine sighting of Marsh Gentians *Gentiana pneumonanthe*. These rarish plants, which occur predominantly in the wet bits of heath, seemed to have benefited from the reduced competition, and the vivid blueness of their belled flowers was set off brilliantly by the charred blackness behind them. These gentians flower late, in August or September, and any site where they occur varies considerably from year to year in how many flowers appear in it. Although they will grow up between the branches of loose heather, sometimes reaching a height of about 30 cm, they flower better in high soil temperatures; so they do tend to flower better in open, unshaded areas.

Another plant which is becoming even less apparent is the Wild Gladiolus *Gladiolus illyricus*. In Britain it can now only be found in a few sites on the New Forest heaths where it was originally discovered by a clergyman in 1856. Not the brash and tarty type of cultivated gladiolus, the Wild Gladiolus has a series of up to eight delicate glowing magenta flowers, laced down one side of the flower spike. Fortunately these vampish ladies are frequently hidden under a canopy of bracken by the time they flower in July and they escape most casual observers. When I arrived to take my photograph two middle-aged ladies had been less casual, and more ignorantly careless, and were brandishing small bunches of gladdies. Such vandalism has hastened its downfall. In recent years the increased number of visitors to the Forest have successfully picked off all but a handful of hidden refugees; and those which are overlooked because they are buried in the bracken are eventually drowned in its enveloping shade. If you are lucky enough to find a colony, admire their beauty and then leave them for others to enjoy.

One plant which has probably benefited from the heather domination of heathland is Common Dodder *Cuscuta epithymum*. Locally called 'Hellweed' or 'The Devil's Guts', it is a parasite of heather, and sometimes gorse, whose brownish-crimson stems vigorously entwine their host. Really it's just a stem which occasionally bears flowers, because it has no roots, and its leaves are reduced to tiny scales. It is a highly specialised stem, however, about 1 mm in diameter, elastic and equipped with modified roots called haustoria which penetrate the host's stem and attach to its vascular tubes. It is from these tubes that it gets all its nutrients, since it has no chlorophyll of its own with which to produce nutrients. From July to September tiny clusters of pinkish-white self-pollinated flowers appear all over the stems, and it looks as if somebody threw spaghetti instead of rice at a wedding.

Marsh Gentian

INVERTEBRATE COMMUNITIES

Because of the poor variety of plants, insects too (many of which depend on the plants for food) have a comparatively low diversity on heathlands. There are only three of our butterflies which are truly associated with the habitat: the Grayling, the Green Hairstreak and the Silver-studded Blue. Many more species can be found, but they are only visitors; and in fact the Large and Small Heath butterflies are misnamed since they are more typical of grassland habitats. Of the three the Grayling, which is on the wing from July to September, and the Green Hairstreak (our only green butterfly) are not restricted to heathlands because their larval food plants are grasses and broom respectively. But the soft growing tips of Ling, Bell Heather and the gorses provide the only larval food of the Silver-studded Blue, and consequently this species's distribution closely matches that of the lowland heaths throughout its range. On sunny mornings in July clouds of these butterflies are active over their colonies, myriads of powder blue males in competition for brown mates. Females emerge from their pupation as the sun warms up their ant-nest hotels, where the pupae have been cared for and overwintered. On emergence they pump up and dry out their wings before releasing a mating pheromone. The result is pandemonium. Males swarm in avid search of them and several may get to a single female at once. She may fly off but is soon caught by these dainty dandies. All tumble to the ground and in a confused whirring of wings somehow one of the males manages to attach himself to the flabbergasted female's abdomen. Quite how it is decided just which male scores in the little twinkling mass of butterflies I couldn't suggest. The united pair continue to be harried by a few persistent Romeos, but this soon subsides. When they part company the female is left alone to prepare to lay her eggs. If you want to catch this action you'll have to be on the heath early because the females can only last a few minutes *virgo intacta*.

The spider community is the one exception to the species-poor reputation of heathland. In fact, heathland boasts the richest diversity of any northern European habitat, with some two hundred and fifty species being directly associated with it. As carnivores, spiders are not directly dependent on the heather and gorse for food, but the obvious burning question is: how do so many different species of one group of animals all manage to eke out an existence in a habitat so poor from practically every other point of view? The answer is niche separation, developed by a need to avoid competition for food and space with other spiders. The spiders have three methods of achieving this spacing of life-styles.

Firstly, different species appear at different stages of the habitat's vegetational development. Very different sets of spider species are found on areas of newly turned heath, and through all the stages of the heather and gorse life-cycle, up to mature heath.

Secondly, because both heather and gorse are evergreen and such permanent solid structures, different spider species become adult (and thus hungry) at different times of the year.

Lastly, within that structure, niche stratification takes place: that is, different species adapt to live under the litter, in the litter, on the litter, at the base of the heather, half way up the heather, and so on to the top of the gorse bush. Using all these parameters, the spiders manage to spread themselves successfully over a spectrum of space and time and live in a kind of dynamic harmony. New species evolve, others become extinct. It happens all the time, everywhere, to everything. It's why life forms are so diverse – because of a need to be different in order to survive.

RIGHT **A male Silver-studded Blue**

LADYBIRD SPIDER

In the summer of 1979, while undertaking some routine fieldwork, a team of scientists from Furzebrook Research Station in the Dorset heathland stumbled upon an amazing rediscovery. In one of their insect traps was a dead male Ladybird Spider *Eresus niger*. This glamorous little gem had not been seen in England since 1906 and was known only from a handful of specimens found between its discovery in 1816 and this date. In northern Europe it is equally rare, although it may be found as far north as Denmark. The males with their splendid orange abdomens marked with three pairs of black spots (which give them their name) are only active as adults for a few days, so it was not until 1980 that Dr Peter Merrett, one of our leading arachnid specialists, could hope to look for a living specimen on the heath. It must have been a long winter. His official rediscovery of the species in May 1980 on a tiny piece of heathland may not have had the scale of adventure seen in Indiana Jones, and Peter isn't a Harrison Ford, but it surely had all the romance of Conan Doyle or Rider Haggard, and in that little lost world he must have been afire with excitement when he picked up his first male *Eresus* – even if he did contain it in his reserved English nature.

Unfortunately, a plague of Heather Beetles attacked this tender little site, defoliating the heather and driving away many of the spiders' insect prey. The spiders starved and their numbers crashed, so that in 1982 there remained only five females and two males. I saw Peter on one of the first hot days of 1986, and he reported that he had seen ten females and twelve males that spring. It seems that inbreeding and lack of space are not a problem for these spiders, since they disperse over very short distances and viable colonies can last for tens of years on a few

3. THE LADY, THE CANDY KILLER AND THE RAGAMUFFIN

LEFT **A patient *Thomisus onustus***

square metres of heath. So even in this age of overknown, overtrodden and overgrown heaths it is possible that more colonies could still remain overlooked. Perhaps on a small sunny scrape somewhere in Surrey, Dorset or Hampshire there are more clandestine families of these flamboyant fellows hiding out.

One day in late June, I went with Peter to look for another spider which is restricted to our heaths. It was one of those picture-postcard days of azure sky, puffy white cirrus clouds and mauve-tinged hummocks spangled with ochre gorses. Hiding somewhere on the many millions of flowers of the Cross-leaved Heath was our quarry – the pink Crab Spider *Thomisus onustus*. It was clear that we desperately needed a 'search image'. That is some ethological/optical/mental acquired adaptation to allow optimal perception off our cryptic prey! (Go and look it up.) All that is commonly known as 'getting one's eye in' to birders and thieves alike. Peter, I hoped, had this adaptation. I had to be content to pursue a course of manic determination. It worked! After an hour of refusing the distraction of the bevy of Silver-studded Blues, Bog Asphodels and 'seep'ing pipits, we found a female. Well . . . OK . . . Peter found a female. He had detected it because it had already killed a Honey Bee worker which it held drooping from its chelicerae (fangs), sucking at it like an overmade-up tart with a Harvey Wallbanger. It was spotted because the bee was held upside-down. Now bees seldom remain static, let alone upside-down on Cross-leaved Heath, so frankly the tiny scene was a dead giveaway (*sic*).

The gravid mother spider was superb: beautiful, a crescendo of crypsis. Her body was not quite the polished pure pink of the heath flowers, more a sunburned-old-lady pink punctuated by delicate cream spots along her limbs. A piece of crazy, mouldy candy – bizarre bubblegum. So sweet, so brilliantly designed to kill.

Her camouflage allows her to sit on what would otherwise be a stupidly exposed position atop the flowers, waiting in patient ambush for pollinating insects. When such prey arrives she begins to dance. In jerky steps she moves her open forelegs to arch around the victim as if to steer it gently into line with her head, so patiently, so precisely and seemingly so nonchalantly. There is no savage battle. She grabs the unsuspecting guest behind the head, snaps her legs closed, and it is dead, or at least paralysed into the pathetic, in seconds. About three seconds, I'd say.

Although the chelicerae are comparatively small they hold a very potent poison, and this enables these spiders to take on insects as large and dangerous as bumblebees. After a few seconds she opens her legs again and the corpse is hung as it is sucked dry. It is thus not bound up and mutilated like a webspinner's prey is; it merely becomes a dry shell to be discarded on a summer breeze after a few hours. The spider swells, irons out her wrinkled flesh and then settles in stillness to wait once more. It must be said, however, that this little *de Medici* often appears frustratingly tired and clumsy. Unless the prey is aligned properly the spider's crablike, sideways, spasmodic and slow cavortings are useless, and her potential victim forages both unaware and unattacked.

The males of the species are sadly shrivelled, warty little brown jobs, bordering on the uninspiring. Peter found two of these 3 mm sex machines – such is the search image of a field scientist of his calibre. Despite perverse encouragement, we could not stimulate any mating behaviour. Normally the courtship is a case of touch and go: despite their good eyesight it is a tactile affair, the male tickling the abdomen and carapace of the female before handing over his half of the genes. Handing is the operative word, since, as in all spiders, his intromittent organs are situated on his palps and not in connection with his testes on the underside of his abdomen. Thus he transfers sperm from

Pillows of gorse

his testes to hollows in his palps where it is stored until he passes it to the female as he hangs tenuously beneath her.

DARTFORD WARBLER

A partial by-product of the super-spiderisation of the heathland is the presence of Britain's single true heathland bird, the Dartford Warbler. Perhaps I'm a little blasé, but I must say they're not my favourites. They always look so scruffy. Dark, skulky little ragamuffins that skip low over the heather when disturbed, their long graduated tails making them readily identifiable as they slip up onto their precious gorse, flicking their flipping tails. They often look as if they've just been pulled through it! As warblers they lack the charm and elegance of their Wood, Garden, and Blackcap cousins. They're Hampshire 'mushes', Dorset 'nippers', with tatty crests, spotty chins and bloodshot eyes. Their calls are grating metallic 'tchurrs' and 'tucs' and once heard are never forgotten. On dismal days you can hear them far off in the brown clouds of furze, then it'll take you hours to get a really good view, unless they have young, when you shouldn't be looking anyway.

It now occurs on many southern English heaths, but its strongholds are in Dorset and the New Forest. Principally a bird of the southern European *maquis*, it occurs sparingly on the northern French heaths but is absent from those further northwards on the continent. So on the edge of its range in Britain and France, heathland is the nearest structurally and climatically similar habitat for this species to occupy. It survives here through an intense relationship with gorse which exerts a profound effect on its feeding and breeding ecology.

Strangely, for nesting purposes the warblers prefer the comfort of old heather and not the gorse where they spend up to eighty per cent of their foraging time. However, birds with more gorse in their territories tend to breed earlier in the year and increase productivity by rearing another brood later in the summer, so-called double clutching. This is a good policy because they have a small clutch size compared with most other warblers, laying only an average of four eggs, and because they also experience a higher infertility rate than their congeners (this may be due to the lower productivity of heathland). They still, however, manage a higher breeding success than the rest of the warblers, perhaps due to the scarcity of mammalian and avian predators on the heath. Adders and Sand Lizards may take a few nestlings in the first few days after hatching when they are especially noisy, but such predators are negligible compared with the number and diversity of those found in the woods and hedgerows. Because of the evergreen nature and structure of gorse the warblers' wholly insectivorous diet can be sustained all year round and this is why the Dartford is only a partial migrant, while its fellow warblers desert the barren winter woods for African climes and foods. This partial migrancy is probably the key to its continued presence: the species is particularly susceptible to prolonged cold winter weather, and the few birds returning from further south are believed to supplement and speed up its recovery and expansion after any dramatic collapse in numbers.

In the eighteenth century there were some 40,000 ha. of heathlands in Dorset, divided into ten large blocks by river valleys. In 1980 only 17,000 ha. remained. The rest had been afforested, reclaimed for agriculture and, more recently, squashed under a horrific urban sprawl. Worse, today a map of the distribution of those remaining heaths resembles a fine spotty mass of printing errors. Tiny, minute even, dots – not those wonderful, big, black ink-splodges of two centuries ago. There are about 768 fragments, of which only fourteen are larger than 100 ha., and sixty per cent of which are less than 1 ha. in area. Such fragmentation has occurred to the heaths all over northern Europe where the scraps remaining are now islands isolated in seas of other habitats; the effect this has on their ecology is traumatic.

Firstly, the number of species that can survive on an island is related to the island, so as the area of the island increases so does the number of species present. This 'species richness' is also the product of a balance between immigration and extinction. Obviously islands that are near to other islands will receive more immigrants dispersing from these neighbours than those which are distant and isolated. Hence for a high species diversity the ideal is either no islands at all, just one huge tract of the same habitat, or at least large islands which are close together. Basically we don't have either situation. In fact we have the reverse, hundreds of tiny fragments well spaced in an abyss of forestry, agriculture and urbanisation – ecological anarchy.

And this isn't the only problem. Because our remaining bits of heath are not real islands contained by inert beaches and sea, they suffer from 'edge effects' or 'spillover' of alien species from their surrounding habitats, and this effect is accentuated because heathland is such a

LEFT **Adders; dangerous when provoked, harmless when respected**
RIGHT **Heathland destruction**

Harebells

species-poor community. Now what happens is that the heathland fragment eventually becomes no more than part of a habitat mosaic and not really heathland at all: it contains so many scrub, woodland and aquatic species that the real 'heath' loses its original identity.

In this age of missiles not morals, profit not posterity, we are then faced with a decision on how best to preserve our 'heaths'. In the light of the biological facts given above, do we, with our limited funds, go for large, pure heathland reserves with their reduced edge effects and healthy populations of 'real' heathland species, or for clumps of small heathland reserves which suffer edge effects, but which will attract more species overall? Since heathland is a seral community in a dynamic state of change, which if left to itself will turn into pine or birch woodland, management is needed regularly to arrest development to this stage and maintain its true identity. This would clearly be more

practical on larger reserves where the several stages of the vegetational succession could be cyclically maintained to increase the diversity and stability of the habitat. Because edge effects are so influential it may also be important to choose areas of heath with vegetation types surrounding them which are complementary, having already reduced these effects by choosing areas greater than 10 ha. in area.

Species with a well-developed dispersal capability such as the Dartford Warbler will suffer eventually from fragmentation, since the young who move off at the end of the summer to find territories and found new colonies, will get lost in the 'seas' of coniferous trees or drown in gaudy fields of rape, never finding their tiny islands in the sun. Large reserves certainly seem best suited to preserve this habitat as a whole, but when rare species only occur on small established sites, these will need considering, especially if, like the Ladybird Spider, they have had little or no dispersal

After the fire: charred heather and gorse

capability and would thus not colonise any large reserves that are set up in the future.

In the past things were effectively kept in check by the heathland way of life, the grazing of sheep, the furze cutters, the peat diggers, and the ubiquitous rabbits. Nylon for 'woolies', coal, oil and gas for fuel, and myxomatosis put an end to this system long ago, and increased forestry added to the speed at which the heath matured into scrubs and woodlands.

There is a magnificent misconception that fire has always been some natural force in the maintenance of heathland, as it is on the plains of Africa and the scrubs of Western Australia. Lightning strikes ye olde pine on the hillock and the fourteenth-century shepherds watch the whole lot go up. Abracadabra, brand-new heath. Nonsense. Fire was only introduced as a management 'technique' in the 1800s to control the encroaching bracken and birch because the decreasing number of sheep couldn't keep it at bay. Resorting to fire is far from ideal, since it can easily lead to nutrient starvation: eighty per cent of the precious nitrogen goes up in smoke, the ash is blown away and the soil is leached and eroded. Badly managed burning can actually lead to the encouragement of scrub and bracken. Carefully controlled burns are now necessary for maintenance, but the tiny fragments can no longer tolerate hot holocausts because they will lead to the local extinction of the species with low dispersal abilities that live there.

Just such a species is the Sand Lizard which in England has, following fragmentation and fire, now declined to a precarious population of perhaps 5000 adults. If the lizard survives the flames, what are his options? To return to the exposed heath to be Kestrel food, move to suburban rockery to be cat food, or starve in the forests and fields? Even if the lizard does survive, and the heathland did begin to regrow, he would never find his way back again. Luckily as yet some heaths are unburned. . . .

SAND LIZARD

Between seven and eight o'clock in the morning in the early days of May, male Sand Lizards fight over females with ferocious enthusiasm. I've watched these Fabergé reptilians from the dappled shade of newly leafed birch on the remnants of the Dorset heaths. This is their stronghold in England, although isolated populations are found in Surrey and Lancashire. In Europe they occur in similar habitats northwards to southern Scandinavia, where they are now specially protected. They are difficult to see anywhere unless you go specifically to look for them because of their cryptic colouration, typical of so many heathland animals. These resplendent little knights in their emerald armoury, bespeckled with ocellate jewels, are heavier, stockier, shorter legged relatives of the viviparous lizard, with a dark dorsal stripe and a bulkier head. In isolation their brilliant finery is bold and beautiful, but when they bask in the top of a bouquet of heather in the half shade of bracken and gorse they become irritatingly obscure.

In spring the males emerge first from hibernation so they can lounge in the yellow sun to complete their spermatogenesis and whip up a tangy green tan. It was thought that they were territorial, but any jostling with other males in fact involves a social, size hierarchy resulting in a displacement of lesser males, thus superficially resembling territoriality. A lizard watches from aloft in the heather, and if another male of any consequence appears he tumbles down for a bit of rough stuff. All animals are loath actually to fight because of the expense of time and energy involved and the obvious risk of suffering an impairing injury. Consequently, ritualised threatening behaviour patterns have developed to avoid any conflict long before the blood flows. Only if all of these preliminary displays, signals and releasers fail will two similarly matched and motivated males actually fight. To see such an encounter is a rare and exciting event.

I found a large male basking 'legs up' in a bush of mature heather on the south side of a tumulus earthwork – prime habitat. He was a striking old rogue in full battle dress; he cocked his head and pierced my stealth with his seemingly knowing eye. He was very special, somehow like the definitive text-book Sand Lizard, and he was doomed. In the shade of the birches his home had outgrown him. Ignorant of this he was adamant, bolshy and a snob. As I settled, another male had, unbeknown to me, arrived on the tawny sand at the base of a clump some two metres away. My lizard supreme, in the green corner, tumbled from his throne and scuttled into the arena in great haste. Luckily for me the combatants had met in a fairly exposed stadium. On approaching each other, both males rose on their forelegs, arching their heads down, stretching their flanks to maximise their greenishness. They both seemed to swell in splendour and size. Normally, hissing by the defender and wide mouth gaping are seen before the butting of flanks acts as the final precursor to violence. At either of these stages the loser will usually break off and be chased away by the victor. But this contest escalated rapidly. With a flippant gasp and a passive hiss my prince exploded into his usurper, biting savagely at his head, neck and jaws, and flicking relentlessly with his flexing torso. The two wrestlers wriggled, their grips regripped, they spun in the dust. They hated, they wanted, they were blinded by fear, driven by adrenalin and fired by the force of survival. It was about the savage selection of the stronger to breed, about a bundle of replicating protein as old as life itself. Genes and Evolution. Darwinism in action.

The usurper failed, he broke off and fled, more falling down the bank than fleeing, hotly pursued by the glorious old brute. The victor gave up pursuit and through my binoculars I watched him as he

RIGHT **A male Sand Lizard in breeding finery**

The symmetry of a sunset in a heathland pool

licked his lips for a minute before he skulked away out of the sun, and the incident died as if it had never happened. I wondered how many times he had fought, how many times he had lost before he had learned to win, how many lizardlings he had sired as Lizard Lord. As my world expanded again I saw a JCB juddering in the haze of a housing estate, and thought I'd go home for breakfast.

Sand Lizard is apt, since these lizards have a profound affinity for warm, sandy, south-facing slopes with deep, bushy heather. They are colonial, and may communally rest in burrows, which are laboriously excavated with their forelegs and head, which they use as a pick. Shorter burrows are also dug by the females for the deposition of their eggs. These they bury well away from any shade so as to maximise the heat from the sun as an incubator. They are entirely insectivorous, feeding mainly on beetles, spiders and flies, and they have a stomach which is much larger than is required to hold their daily intake of food. This suggests that it

may have developed so that they can exploit locally and seasonally abundant food sources by veritably stuffing themselves when these are encountered.

SMOOTH SNAKE

This is another reptile rarity which is even harder to find in the tangle of Europe's heathlands. The Smooth Snake I find a bit of a bore. It has neither the panache and looks of the Adder nor the boyish charm of the lovely Grass Snake. I once caught a huge specimen, at the time one of the largest on record, and I was still mainly impressed with its rarity and novelty. I stumbled across this insane individual right out in the open, fast asleep and neatly coiled, like a cross between a raffia placemat and a squashed pony-dung. On the pretext of caring for its safety I woke it up, and it bit me – because they do bite frequently, their sharp little rows of teeth drawing blood from painless wounds!

Not even discovered in Britain until 1859, Smooth Snakes were once so common on the slopes around the village of

A Smooth Snake on *Sphagnum*

Bournemouth that in the 'hot summer of 1866 literally scores were killed'. In these 'large expanses of moorland intersected with valley bogs . . . which were a favourite hunting [*sic*] ground for naturalist . . . their numbers gradually decreased'. The decline has continued unabated. Bell and Brusher Mills, Victorian snakers of repute, found very few in their otherwise glorious days of snakery. This was certainly due not only to rarity but also the species's incredible invisibility. They are more subterranean than the Sand Lizard and, even if they are not hibernating, which they do for half the year, they are doing nothing at all under all the litter at the base of some old heather. A friend of mine, Tony Gent, has studied these enthralling beasts and has found that they usually move about two metres an hour, if they move at all, and he calls them cryptic heliotherms (among other things). He means that they would like to keep their body temperature at about 30°C, but in the wild it is usually at about 23°C. This is because they are loath to expose themselves by basking to raise their temperature. Thus they settle for a trade-off, perhaps showing themselves for a brief period in the base of a gap between two heather bushes and then slothing around in the dark. This reduced activity means that they feed very rarely on their evenly mixed diet of other reptiles and small mammal nestlings. Tony concluded that they might eat a maximum of six lizards a year, and such a limited food intake results in the females only producing young biennially or triennially, and living a long life. These snakes share the same habitat preferences as Sand Lizards in Britain, while in northern Europe they frequent dry banks, scrub and even hedgerows. In such sites up to twenty adults per hectare can be found, and estimates for the overall British population size range from 1,000 to 52,000, but about 7,000 to 10,000 seems a more reasonable assessment for the present total. Who knows exactly, and who will ever know, I wonder, since this animal is so secretive and seems immune to practised census techniques?

If you are asked what the most ferocious animal in the world is, you tend to think in terms of sharks, grizzly bears, crocodiles or tigers. You always think from the human point of view. For me some of the most fearsome animals of all live out in these blasted wastelands. Solitary soldiers with an Attilan mentality, and the same crisp, cold, precise attitude to the ultimate sin as a Samurai swordsman. These aren't vandals like a Badger in a Bees' nest, or Crows pecking the eyes out of lame lambs. These are assassins. Devastatingly effective. Relentless. Merciless. Armed with scimitars and stings, they kill; and that's it.

TIGER BEETLE

A Tiger with no stripes. This beetle was named after a rabid Mandalay man-eater with a desperate headache. He is the northern scout of the *Cicindela* warriors. He runs at great speed over the sand on long legs of brilliant burnished copper. He bursts into short flight, his elytra flashing like magpie's wings, like iridescent Athenian shields in the striking sunshine. Catch him, carefully hold him face to your face, look at his scythe-like mandibles crossed horizontally, now flicking as he flinches and imagine you are an invertebrate. Fear and ferocity!

Cicindela campestris has prominent eyes and appears clad in a range of metallic colours ranging from black to brilliant green, always hemmed with a metallic seam on his backplates. He kills on the ground, his children kill from underground. He chases, they wait in burrows, tacked in with stiff spines to a chamber perhaps 30 cm deep. An insect passes and they shoot out to grab it with their equally exaggerated mandibles, and then return to the depths to devour it. Here they pupate, to emerge in full fury in the first warm days of summer. In flight the Tiger Beetle is rapid, direct and fairly talentless. His co-killers on the heath are accomplished.

LEFT **The ferocious face of the Tiger Beetle**
RIGHT **A juicy morsel**

SAND-LOVER WASP

Although lacking the dexterity and the airborne precision of the dragonflies, the Sand-Lover Wasp *Ammophila sabulosa* typifies menace in an insect guise. Find yourself a sandy bank in June or July in the lee of some vegetation, sit down for a tan and watch this macabre military at work. In the air their flight has a dream or nightmare quality. Wafting, easy, both slow and stupidly fast, always deliberate and definitely dangerous. On the ground they transform into a totally manic murdering machine. Insane with ceaseless energy, twisting, turning, running in a psychopathic frenzy to search out and kill. Quite horrible!

While the Tiger Beetle kills to satisfy his own appetite, these beasts paralyse caterpillars for the benefit of their young. Indeed, if the adults do feed at all they gently sip nectar from shallow flowers. The young are provisioned in burrows and if you're lucky you may see an adult dig such a chamber. The female chooses a spot with some firm sand and begins to dig with her forelimbs a chamber about five centimetres deep, in bursts of feverish activity. They can be very closely approached, they become so preoccupied with their tasks, be it digging or killing. The movement becomes robotic and increasingly diligent. If larger particles are encountered they are removed in the mandibles, the wasp appearing to fly backwards out of its hole to ditch them a short distance away. One of the discarded lumps may later be retrieved to block the entrance. The female then jumps for joy at the completion of her task, hopping, running, flying about on the sand. In fact she is learning the whereabouts of her burrow, using any vegetation or distinctive features as guides to its exact location. She may travel many metres in search of prey, or even rest somewhere else overnight; so it is important that she can find her way back to this and to several other chambers she has made.

Her search for prey is feverish and on foot. Darting about on the sand beneath the vegetation, every hiding place is systematically searched by sight, and probably by smell. Backwards and forwards, over and under, until it intensifies upon a particular cleft in the base of the grass or gorse. The kill occurs in another dimension. Gone is the panic, the manic marauding and the menace. The caterpillar is extracted in what appears now as slow motion. It is juggled until the murderess holds its head in her jaws. Then with diabolical deliberation she stabs it several times down its torso with the tip of her abdomen. It's so clinical, so cruel.

The fatally crippled caterpillar is dragged back to the burrow where she lays an egg on it. On hatching, the larva begins to eat it alive, leaving the vital organs until last. She returns to these calculating murderers, bringing them yet more poor victims, until she is satisfied that they are fully grown; she then entombs them with their brothers and sisters before she dies. I've seen *Ammophila* die in the wild. One minute they are the typification of everything I've elaborated – then they just die. The fury and ferocity fizzles out in a second, and they get blown across the sand to break up like some old car.

POTTER WASP

Eumenes coarctatus is better known as the Potter Wasp after the delicate little mud nest it fashions on the fronds of heather, and in which it hides the corpses of its prey. Again it provides caterpillars for its young, this time tortricid or geometrid moth larvae which are literally torn out of their cocoons on birch leaves, stung – neuromuscularly raped – before being slung under this black and yellow pirate and packed seven or so at a time into its clay tomb. To make this chamber it collects water from ponds, streams, puddles or dew drops, and flies to a site which has sand of the chosen mix. Here it exudes the water and kneads a pellet of the resulting mud with its mandibles and forelegs. Up to thirty pellets are needed to produce the

RIGHT **A Sand-lover Wasp**

delicate little sphere; within a couple of hours it nears completion, with a nifty little collar turned into the entrance.

Before she flies off to start her pillage she lays a single egg on a thread in the top of the chamber. She then stuffs the larval larder full of caterpillars and, after filling several more nests with eggs and provisions, she dies. Think of it. The wonder of this tiny insect on a bit of heath on this little planet. It's really quite miraculous, the way she's developed this beautiful behaviour; her skill as an artisan and assassin is amazing. Yet the miracle has only just begun because when the adult emerges from the pot it isn't *Eumenes* at all.

CUCKOO WASP

Sitting watching while the Potter was at work was a rubytail, a jewel, a Cuckoo Wasp *Chrysis ignita*. A metallic marvel of evolution, this stunning little cad is a parasite of the potter and she attends the pot with as much interest as the maker – but only in its absence! Eight times she visited the little ceramic capsule in the making, but on the eighth she laid her own egg and then left. Of course, the Potter Wasp later laid its own egg alongside the parasite's, but the catch is simple. The parasite's hatched first, after just a day or so, and guess what the larva's first meal is – the Potter's egg. Then it eats the caterpillars. Then it pupates. If that isn't more amazing than the storyline in *Dallas*, what is?

I have only ever seen two of these chrysid wasps. The first I caught in a net on some sunburned afternoon out on the heath. The second I accidentally collected at night with a spade, and I didn't find it for days. It was dead, dried out and decapitated; yet its discovery was a dazzling spectacle on the microscale.

SUNDEW

The Cuckoo Wasp had been slowly dissected on one of the fizzy paddles of the

Common Sundew *Drosera rotundifolia*. On the wetter areas of the heaths the carnivory continues; but in this example the insects are victims of plants.

Because of the very poor soil, starved of nitrogen, these pretty little plants have developed a process of passive predation to supplement their nutrient intake. Their leaves are arranged in a radial rosette on short stalks, each of which supports a highly specialised blade. This is covered with a mass of red tentacles and each of these is tipped with a small egg-shaped gland which secretes sticky mucilage to trap the insect. It then secretes the protein-dissolving enzymes, and lastly actively imbibes the resulting soup, leaving only the chitinous shell of the victim. Why insects land on these traps in the first place is unknown. The dewy drops certainly look nectarous, and it is likely that they smell so as well. One plant can account for up to two hundred insects through the summer, from hapless little midges to our poor chrysid wasps and even damselflies.

Once stuck, the struggling insect stimulates the tentacles by touch, and they begin their slow constrictive caress. Each tentacle begins to grow, the outside more than the inner, resulting in a curving into the leaf centre. Here the prey is tightly compressed, the leaf itself buckling up with larger prey to bring more glands into contact with the surfaces of the now long-expired victim. Once the digestion is over, the tentacles grow with a bias to the inside and they straighten out. There is a limit to the total growth of the tentacle, so this process is only completed about three times before the leaf is exhausted and withers.

In the past these plants were valued for their medicinal virtues, notably for killing people already dying of consumption, for an exceedingly alcoholic liquor, for removing corns or warts; or, when mixed with milk, they were used to remove freckles. On the farm, however, sundews were considered a nuisance: 'Sheep and other cattel if they do onlye taste of it are provoked to lust.'

LEFT **Fizzy petals of sundew**

39

Late summer, and the heath is a field of honey and smells as sweet. Millions of bees buzz over the burgundy heath caressing each tiny flower again and again until they have all been bled of their nectar and have been fully fertilised. The sun's gone out and the sky is a rich grey with a dash of purple like a reflection off the wasteland. A strip of anaemic white separates the two on the horizon and it feels sticky. It's just so sickly thick that your head is heavy with the perfume, aching with the air and bumping with the bee buzz.

He'll appear like the ambassador of storm on wings the colour of thunder. With an energy way out of line on a day like this. He's low, a silhouette some three metres over the blurring ground, slicing the air like a psychopathic swift. He's in control, using a vibrant bouncing flight to twist and turn. In a spasm he'll stall and turn and flip up with his legs out to catch an invisible insect. Then he'll slow up and crease up neck down and feet up in a glide, and you'll see some tiny wings reeling away like sycamore seeds. As he nears he'll repeat this several times with a near effortlessness. Somehow you can sense he's only on half speed until a pipit flicks and he follows with turbo. The rhythm goes. Danger explodes, and speed. They twist and turn and now you see his colour. He's red above his feet as if he has perched on a 'phone box ignoring a 'wet paint' sign. He's bespeckled black on cream and has a large moustache like Groucho Marx. The chase tightens and heads skywards. They fuse, but then the cohesion slips and they become two again. The pipit drops, the pursuer pursues but now it's in pretend. He's lost the chase, because although he follows his prey down to the safety of the tangled purple he could never catch him terrestrially. He soars up and within seconds he has another insect in another

LEFT **Onset of winter at a heathland pool**
RIGHT **Marsh Helleborine**

speed. Then he turns nonchalantly into the white strip of sky and disappears like the dot on a TV screen in negative. He came and played a killing dance for you for a minute, but you can only remember a second or two, a flection of his wings, a colour, a movement off his mind in yours. It's like the memory of a nightmare. A storm strums from behind some distant pines.

THE HOBBY
The elegance and aerial agility of this little falcon can only be understated. You must see it to feel it, and to see it in Britain you'll have to go south of the Thames to be sure. On the continent Hobbies are never common, especially further north; they have suffered marked declines in France and Denmark, while in Holland they strangely appear to be holding their own, or even increasing. This decline is probably nowadays habitat-related, although in the past they must have suffered from pesticides' evil infusion, and in Britain at least they were at times locally hammered by the exploits of egg-collectors who prized their rich tanned shells for their seedy, dark, moth-ridden cabinets. Hobbies are principally bird feeders, doing especially well on the bevy of young swallows and martins which usually coincide with the period of having young in their own nest at the end of summer. Despite this, to see them your best bet is to look out over the bogs and mires as they hawk for insects earlier in the year. Indeed, these damp valleys are some of the most interesting areas of our heathlands, richest in plants, and home for another whole school of aerial predators – the dragonflies.

DRAGONFLIES
Dragonfly science is, as far as the amateur naturalist is concerned, a new one. I suppose it was considered too difficult to identify this array of fast-moving aeronauts, and it was only in 1983 that the British Dragonfly Society was set up. Since then a healthy supply of field guides has made the European *Odonata* readily identifiable.

Confusion still reigns among the common names, with the genera *Orthetrum*, *Sympetrum* and *Aeshna* sometimes being referred to as skimmers, darters and hawkers respectively, and things like the green *Lestes* genus being the one and the same as the emerald damselflies.

For this reason the 'Latin' (scientific) names still find frequent and necessary use, and in fact they really are worth learning for all of the flora and fauna, especially if you are an English person abroad. If you can't ask for a box of matches in Spanish, the way to a Hungarian railway station, or for the time in Swedish, you can at least make any respectable naturalist in a foreign quagmire understand your problems with the good old scientific name!

Back to the dragonflies. Well, just like many of the other heathland animals and plants many dragonflies are also on the edge of their ranges in northern Europe and especially in Britain, and yes – you've got it now – they occur on heathland because their thermal requirements make this the most sympathetic habitat structure.

Large and varied heathland bogs, flushes and streams are of exceptional value to dragonflies, especially if there is a range of habitat adjacent to the heath and it is not too acid. Such areas often attract cosmopolitan species such as *Libellula fulva* (the Scarce Chaser), *Lestes sponsa* (the Emerald Damselfly) and *Enallagma cyathigerum* (the Common Blue Damselfly); but some occur because they can tolerate the acid, such as *Libellula quadrimaculata* (the Four-spotted Chaser), *Ceriagrion tenellum* (the Small Red Damselfly) and *Aeshna juncea* (the Common Hawker). Other species are ecologically entwined with the bogs. *Leucorrhinia dubia* (the White-faced Darter), a rare British species occurring principally on the Surrey heaths, needs the shallows

RIGHT **A damselfly, victim of a sundew**

A dewy dragonfly *Libellula depressa*, at dawn

of boggy pools to lay its eggs. The White-faced Darter sometimes lays directly onto wet *Sphagnum*, but *Aesna subartica* (the Subarctic Hawker, a European species very similar to *Aeshna juncea*), has this as a precise requirement. *Aeshna caerulea* (the Azure Hawker) also shares this habit which must stem from the yet undiscovered requirements of their larval stages.

The beautiful greeny and metallic *Calopteryx virgo* (the Beautiful Demoiselle) occurs along flowing heathland streams in the New Forest where its cousin the rare *Coenagrion mercuriale* (the Southern Damselfly) frequents the banks, favouring areas of bog myrtle to rest on. The New Forest's two rare damselflies are *Ceriagrion tenellum* (the Small Red Damselfly) which has a weak, fluttering flight and often succumbs to sundews in the bogs it frequents, and *Ischnura pumilio* (the Scarce Blue-tailed Damselfly), which also has a weak flight with frequent rests.

Dragonflies in the New Forest include: *Orthetrum coerulescens* (the Keeled Skimmer) which in its favoured *Sphagnum* areas may be very abundant at times; *Sympetrum danae* (the Black Darter), which is a non-territorial species often occurring in masses, especially in the north where it is more frequent; *Aeshna cyanea* (the Southern Hawker), a beautiful yet common jewel which is often inquisitive, and sometimes flies on into the night and in colder, duller weather; *Aeshna mixta* (the Migrant Hawker), *Cordulegaster boltonii* (the Golden-ringed Dragonfly) and *Anax imperator* (the Emperor Dragonfly), all of which are larger dashing species which patrol the mires hawking, and being hawked, in the summer sunshine.

RAFT SPIDERS

Another bog-bound favourite, *Dolomedes fimbriatus*, the Raft Spider, I had never seen before, and I wanted to photograph it for this book. A friend told me of a location

A Raft Spider, lying in wait underwater

in the New Forest; so on a brilliantly hot July afternoon I wandered out to it through a monotonous Sitka Spruce plantation, which I can only say smelled nice. The pool had survived a forested fate because if was in an area too wet to plant. It was idyllic, better than any Pre-Raphaelite Ophelia or Lady of Shallot scene. It had these colours: deep, rich and intense greens, yellows, reds and blacks. This little oasis was surrounded by heather and marram grass, was very deep and jet glassy black, yet the only clear water was a central, circular area of about 2 metres in diameter. The rest was a bed of water-lilies in flower. They were not, I think, our native White Water-lily *Nymphaea alba*, but a horticultural cultivar which had escaped to pollute the countryside. But I was in no mood for environmental snobbery or racism. They were cliché-beautiful. Petals of veined porcelain, a dreamy white of imperceptible tone and light. I lay in the mud to photograph a back-lit flower. Were the petals that were

shaded by others grey? Not really, yet they were of a lesser light than the lit ones, which were more than white. The sexual parts of the plant were a frosty canary yellow to ochre, depending on the incidence of the sun, looking like delicate marzipans dipped in lemon sherbet. I failed to transmit any of this to Kodachrome. Photography is not for the romantic, more for the cruelty of the realist. Oh yes, what about the spiders?

I lifted up a few lily leaves around the side of the pond, and there they were. Totally amazing. There were females everywhere; the pond was alive with them. Heaven knows what they were all eating – there must have been even more of that. Raft Spiders are of course voracious carnivores; indeed, one even bit me. It surprised me, but didn't hurt. I was enraptured by their defiance, their wetness and wondrous lifestyle. I put one in a jar and forced it to hide under a fragment of lily leaf. It stuck inverted like a spider-shaped bubble to the leaf. A single eight-legged bubble of mercury, and when it moved it moved with the same repulsive freedom of that mad metal.

These are large spiders, the females sometimes reaching over 2 cm from head to tip of abdomen, with a leg span of up to 6–7 cm. They are a dark olivey-brown with two yellowy-cream stripes (sub-marginal bands) which run from face to rear on the abdomen. They appear very sturdy and threatening when they perch on their lily pirate ships. They can skate across the water surface either to escape or to capture their food, which they probably detect either by the vibrations it produces or with their keen eyesight. Large prey is taken, bluebottles, horseflies, and damselflies, some of which are scavenged once they have fallen into the water and are struggling on its surface, and the others captured by pouncing. I suppose they are most famous for their ability to catch small fish, mainly minnowlings, which they attract to the surface by vibrating the water with their front legs. The tiny fish are naïvely curious and, as they gather and begin to nibble the spider's ripples, it pounces pin sharp and seizes its victim. This it then holds in the meniscus and, as it sucks and stabs, it occasionally repositions its grip to ensure that all of the flesh is finally mixed up and dissolved by its digestive juices. In Britain they are locally common in the south although absent further northwards, being found in early summer (adults) on areas of permanent water. In Europe they are widespread.

THE FLORA OF BOGS AND MIRES

As remarked previously the flora of the dry heath is poor; but, as the soil wetness increases down into the valley bogs and mires, so does the diversity and interest of the flora.

In places the Cross-leaved Heath grows with the Ling and Deergrass – Purple Moor-grass and Cotton-grass taking over in areas of standing water. In autumn these areas of the heath have a glorious golden hue, and seem to glow in the low sunlight, sandwiched between the dull earthy bands of heather. Rarities that grow here include the Marsh Gentian, Marsh Clubmoss *Lycopodiella inundata* and Brown Beak-sedge *Rhynchospora alba*. All three sundews occur, although the Round-leaved is the most common in the south. Bog Asphodel with its marvellous yellow candles of flowers, Heath Spotted Orchid in its many colours, shapes and sizes, Bog Pimpernel, Stagshorn Clubmoss and Common Butterwort can all be discovered one step past the one where your wellingtons fill with the foul-smelling black mud characteristic of these areas! Dorset Heath, Devil's-bit Scabious, Meadow Thistle, Tawny Sedge and Flea Sedge join the bladderworts here, but the most tormenting, sought-after, back-breaking, wellie-filling, knee-aching, botanical prize has to be the most aesthetically pathetic – the Bog Orchid *Hammarbya paludosa*.

I searched high and low in the bogs of the Highlands and Islands, where it is said to be commoner, and across the mires of the New Forest for three summers to see this plant. Eventually a friend discovered a little colony of about thirty-five plants on a steep boggy slope with small patches of *Sphagnum* between Deergrass and open mud. When the Bog Orchid grows in *Sphagnum*, its yellow-green colouring makes it very difficult to see, even when it peeps a few centimetres above the mat of moss. This colony had been discovered because they were growing out of the black mud and were not cryptically hidden, and because of their size. Normally they range between 8 and 10 cm; these stood up to 18 cm tall.

The Bog Orchid is one of the few species which can survive this very acid environment; like many orchids it is entirely dependent on its fungus partner for its nutrition. Its underground or

RIGHT **Bog Bean**

46

underwater parts consist of two tubers surrounded by leafy bracts, and these are often a little distance from the flowering shoot and usually supported in the moss along with its hairlike roots. The leaves are small, ovate and fleshy and come in twos and threes and the flowers are minute greeny-yellow jobs. Their comparatively long stalks are twisted through a full 360° (all of our other orchids are twisted through 180°), so that the lip is pointing downwards and in effect the flower is the correct way up. These flower stalks appear in late summer and early autumn, and it appears that they are pollinated by small flies and midges. Seed pods develop after a successful pollination spree, and the microscopic seeds they discard float on water, giving the plant an effective method of dispersal – in theory at least.

Seeding is not the only method of reproduction, though. Careful examination of the tips of the leaves may reveal a cluster of tiny green buds. It appears that these too are discarded, and can grow into new plants. The species is widespread in Britain and Europe, but, as I have said, you'll get a few wet knees trying to find it.

NATTERJACK TOAD

The inclusion of this toad *Bufo calamita* in a book on heathlands is today, from the British point of view at least, verging on the spurious. It only occurs naturally now on two inland heaths here, although in Europe it still occurs in such habitats, albeit precariously. Principally a western European species, commonest in Spain, it occurs as one of the region's three native bufonids as far north as Denmark, but in all of these northern areas its strongholds are now on the coasts, and it is sufficiently endangered to be classed as 'Vulnerable' by the IUCN Red Data people.

BELOW **A pool of Bog Bean**
RIGHT **Natterjack Toad**

A Natterjack Toad emerging from his sandy burrow

The Natterjack's pronounced affinity for loose, dry sand and open, unshaded, warm areas now means that it survives away from the heath in this habitat, and thankfully sometimes thrives. It is a small toad, up to 8 cm long, stocky and short-legged, typically having a characteristic yellowish stripe down the centre of its back. The rest of its body is an olive-green-ish- greyish-brown with darker mottlings. It seldom jumps or hops, much preferring to run or walk in abrupt bursts over the sand. Natterjacks share the Sand Lizards' tenacity and ability for digging burrows in the soil, and also their taste for mature heather with open sandy banks; it is this habitat structure which they now occupy on the coasts. Dunes at all stages of development, from littoral zones through to the mature fixed structures inland, are occupied, as are sandy islands and sandy fields in parts of Europe. Whether it is on the heath or dune, the all-important breeding pool or pools have precise require-

ments. They need to be shallow, gently sloping to a maximum depth of about 0.5 to 0.7 m, and be totally unshaded. Aquatic vegetation seems undesirable, and the actual size of the pool seems unimportant; but to permit a high percentage of the tadpoles to transform into toadlets this standing water needs to last at least until mid-July and then dry up altogether. This gives the tadpoles enough time to meta-morphose into toadlets and leave the pond before the drying exterminates any of the pond predators, such as dragonfly and water beetle larvae, which would other-wise survive until the next season to prey upon next year's tadpoles.

These toads are nocturnal, being prin-cipally active between dusk and midnight when they feed on their insectivorous diet of flies, beetles, millipedes, ants and bugs. From April the males call from their breed-ing pools, the larger mature individuals calling first and the smaller younger ones later from June onwards. The females visit the pools only briefly, perhaps for twelve hours, and once they have spawned leave immediately. As with many of these southern continentals the Natterjacks' choice of the heathland habitat structure is directed by their thermal requirements, and their decrease in this part of their range is therefore presumably due to habitat loss through man's activities.

However, unlike the Sand Lizard and Smooth Snake, the Natterjack Toad's breeding dynamics make it a more suitable species for active conservation by translocation or reintroduction. A friend of mine, John Buckley, with the encouragement of the RSPB and the British Herpetological Society, has developed just such a project on a 2.5 ha site of English heathland since 1979. Firstly a small shallow pool was kindly dug and lined with plastic sheet by the RAF; then, with all the correct licensing, John moved the equivalent of three spawn strings (to ensure a good genetic mix) containing about 9,000 eggs from an English inland site to the pool in 1980. By the fifth season in 1985 there was a population of 37 breeding females, which laid 71,000 eggs in the pool, and a total population of 80 to 100 adults. Now toads from spawn which had been laid in this pool have returned to breed there, making the population self-supporting. Indeed, the population is already comparable to the two original inland sites in terms of population, and even higher in terms of productivity. This increased productivity was due to predator suppression: water beetles and water boatmen were removed from the pond, as were Common Toads because their tadpoles, which emerge earlier, frequently eat the Natterjacks' spawn. Since this pioneer project's success, two similar projects involving transferring populations to suitable habitats have worked well at a cost of only a few hundred pounds. If this doesn't illustrate what a small sum of money and a lot of well-directed enthusiasm can do for a rare species whose ecology is relatively well understood – what does?

So far all the species mentioned have been visitors to, flowering in, or actively enjoying the summer months on the heath. Why? Well, in winter the heath is pretty horrible, unless you can revel in a wind-worn patch of natural nakedness. It's virtually dead; even the beautiful orange tints of the grasses have gone, and it's browny-grey and flat in colour as well as in topography. Still, despite the demise of the diversity of life, some squatters do arrive for the winter. All birds and all at low densities.

THE GREAT GREY SHRIKE

This striking bird arrives from October onwards, depending on the weather in the northern part of its range, and may return to exactly the same patch or even the same perch as it had occupied the previous winter. In southern England these birds are always unpredictable in number and time of arrival, with peak numbers in January. They are solitary and from one minute being incredibly obvious monochrome, magpie look-alikes, stuck up on familiar perches in all this gloom, can suddenly become quite invisible yet ominously present. You can waste hours scouring every pine sapling in several adjacent valley bogs and their respective ridges, only to have one magically appear alongside you. Their wintering ecology is relatively poorly understood; but from my own observation I should suppose that they exploit very large ranges, but are territorial over much smaller areas inside which they have a patch-foraging strategy. What I mean is that they have a series of places where they are regularly seen hunting, and within these actually hunt from favourite perches which they return to. As such they hunt rather like kestrels (they can also hover like them, sometimes for quite long periods of up to 30 seconds or more), except that because there are few insects on the heath in winter, and equally few small vertebrates, they are principally predators of birds. I presume they stand sentry, spying on the flocks of tits, finches and buntings which traverse the heath occasionally

Clouds of soft Bog-cotton seed heads

between their seed sources, or even await the appearance of one of the home-loving Meadow Pipits. Pursuit is direct, not by surprise, and undertaken over quite long distances. They fly, when shifting perch, very close to the heather with a bouncy finch-like flight, and then whip up at the last minute onto their perch. They sit upright, flicking and flexing their tails in the hideous wind and drizzle, hiding their ferocity behind their pretty little burglar masks. Their curious habit of impaling food is well known, and near one roost I found little bits of small birds all winter in 1982, never whole and all seemingly fresh, pushed into clefts of thorn bushes and gorse. Such behaviour is very suitable for this species, since at this time of year in this place food must come in fits and starts because of the low productivity of the area and consequent scarcity of prey and its unpredictable appearance.

HEN HARRIER

These birds move down from their upland moor breeding-grounds for the winter. Out on the heath they're like tatty rags blowing in gales, flopping upwind just off the ground in a painful search for the only vole on the heath. Tired and lazy, they bounce over the boulders of brown with irregular wingbeats and short glides on high-held wings. Lonely tramps looking for homes far from home, at night they gather in sad roosts where they land without speaking, not even perching near each other, in the deathly shade of afternoon, sharing some sheltered spot like vagrants share a row of park benches. They are anonymous rovers, mute veterans of the rain, soldiers from forgotten wars and widows of the north wind. Less romantically, I suppose it is fair to say their methods and mechanics of flight are lacking in the dash or panache of some of their fellow raptors, and they seem much more distant. I've never had a good look at one perched – I've never seen a harrier's face. I've never even seen a harrier's eye; yet I've peered at the eyes of Peregrines, Kestrels, kites, Hobbies and hawks. I've only ever seen harriers on the other side of open ground in the cold. I must say I don't really know the Hen Harrier at all.

RIGHT **A heathland pool**

Many of the heathland animals are cryptic in appearance. There are individual reasons for this, of course, but many stem from the fact that the habitat is an open one, lacking in much spatial heterogeneity. There is no dense cover to hide the larger species, and the lack of trees means that the ground-nesting birds have to hide themselves and their nests with camouflage. The Sand Lizard is dependent on sunshine to warm it up enough to be active and because of this it needs to bask up in the heather; this exposure has induced its need for camouflage. The Grayling Butterfly and the Crab Spider have to hide from Stonechats and Dartford Warblers, as do the heather look-alike larvae of the Emperor Moth. In many of these animals this crypsis has grown to dominate the species's character, physiology and behavioural ecology.

CAMOUFLAGE

At rest the Nightjar is a bit of tree. Primarily a mottled dusty brown, in close-up it is a very intricate, vermiculated, freckled and barred piece of broken bark. Like so many of its heathland cohabitants it is incredibly well camouflaged. Camouflage is a similarity between the animal and its background, a kind of visual deception by means of which an animal can either elude its predators or lurk unseen awaiting suitable prey. In the case of the Stone Curlew, Sand Lizard, Grayling Butterfly, and the Nightjar, it is the former. *Thomisus*, the Crab Spider, both eludes and ambushes. Animals can be easily detected against a particular background if they have a sharp outline, cast strong shadows or are a different colour. Most camouflage techniques are designed to minimise these differences. Stone Curlews have black and white wing-bars, and the Sand Lizards have contrasting dorsal stripes to break up their body outlines – examples of disruptive colouration.

LEFT **Broken bark hiding on the heath**
RIGHT **A Heathland in the New Forest**

Shadows cast on the ground by resting animals again increase apparency and these are often reduced by a flattened posture as in the roosting Nightjar. The Grayling Butterfly, which by physiological design has to roost with its wings vertical to the body, minimises its shadow by either turning to face the sun head-on or by leaning over on one side as soon as it lands on the gravelly paths it frequents. By doing so, it only casts a thin needle of shadow behind itself or under its wings, and is otherwise obscured by its matching colouration. Indeed, the colour-matching seen in all of the above species is extraordinarily good. Natural selection knows no imperfections: those of a lesser match would soon be eaten out of the breeding population.

For any of these camouflaging methods to work the animal must remain motionless. Indeed, confidence in their crypsis is seen in all these species. One almost has to tread on a Nightjar to flush it, and Graylings explode off the path from under your footsteps. All this contrived invisibility is all very well for hiding from predators, but what about hiding from their own kind? This could make finding a mate difficult, and courtship tedious. Because of such self-imposed restraints sexual selection has led the birds to vocal rather than visual displays; the vociferations of the Stone Curlew and Nightjar are profound examples of this, with their characteristically distinct noisiness. Insects and spiders have a highly developed pheromonal communication with prospective mates, and the Grayling uses this plus an exaggerated display to lure females down onto his path, something not seen in many non-cryptic butterfly species.

At last the heat of the day fades, the flies transform into midges and mosquitoes, the sickly-sweet coconut stench of the gorse is washed away, and you reach for your pullover and begin to scratch. You moan about the vampiric habits of the insects and set off for the last eco-gem of our heaths – the Nightjar.

NIGHTJAR

In a way we've come full circle in our exploration of these wastelands. In the death of daylight I'm going to show you this cryptic, crepuscular, insectivorous, ground-nesting, migrant bird whose habitat is heathland, and – guess what – whose numbers are diminishing rapidly. And it's weird. More like a macabre moth than a high priest of the steppe, the Nightjar is as elusive, and its habitats as relatively unknown, as the Stone Curlew. It is its kind of aerial alter ego.

On arrival in the suitable habitat, woody heathland edge, you will have to wait until about 10 o'clock in the evening in July to hear the soft, nasal whispering of the males as they leave their diurnal roosts to buzz about briefly in the blurred gloom before flying off to feed. Like the Stone Curlews it seems that Nightjars too have separate breeding and feeding areas. Here they gather communally to hawk flying insects, males and females together, up to 5 km from their nest sites.

The Nightjar is a highly adapted specialist feeder: its wide gaping mouth, lined with stiff rictal bristles, is specifically geared for catching insects in flight. Less apparent is its gizzard, which is used as a food-storage organ. It has been noted by bird ringers that, when captured at night, after feeding, these birds have pronounced abdominal bulges. These are their gizzards stuffed with insects. The avian gizzard does not normally function as a food-storage organ, and its development as such in this species suggests that a brief period of maximum prey abundance needs to be exploited to lay down sufficient reserves until the next foraging session. Flying five kilometres to a feeding site is energetically expensive, as is the Nightjar's method of aerial prey capture; so this physiological adaptation seems a realistic solution to an ecological problem and is superficially similar to the method employed by the Sand Lizard.

RIGHT **Marbled Nightjar eggs**

Glowing Bog-cotton

When I used to look at drawings of Nightjars' mouths in my childhood bird-books I used to try to imagine these birds flying round, mouths open, trawling for craneflies and midges all night and wondered how far they'd have to fly to make it worthwhile. Trawling like this doesn't in fact occur, and, although birds have been seen with their bristles caked with midges, this was not the result of individual pursuit, rather just those caught accidentally while pursuing the larger moths and beetles which comprise their diet. Recently, using night vision equipment, observers have seen Nightjars to-ing and fro-ing on short sallies from the same perch, snatching food out of the forest-edge canopy like flycatchers, hawking, gliding about like harriers over the heath, and directly pursuing insects like hobbies do – using their large eyes to find individual items – not mindlessly mouthing for midges at all.

But what of their decline? The explanation may quite simply lie in the fact that, in the past, Nightjars have always been double- brooded, but that today this is exceptional. Females now arrive at their breeding-grounds later, and take longer to lay their first two-egg clutch. In the past the male was left in charge of the young when the chicks were about two weeks old, while the hen went off to lay another pair of eggs. But nowadays even pairs who manage to start early don't bother with another clutch. The growth-rates of the young have been adversely affected by bad weather which, accompanied by less insect food in the recent trend of poor and late springs, may have encouraged delay of the first clutch until the food supply is adequate, at the expense of a second clutch.

Night on the heath

The trouble with this climatic explanation – which is loosely compatible with an explanation for all similar declines in the Stone Curlews', Red-backed Shrikes' and Wrynecks' populations, all migratory insectivores on the edge of their range in Britain and northern Europe – is that the decline set in *before* the climate began to become inhospitable in the early decades of this century.

Habitat loss has been a definite and direct blow, and so has the use of pesticides. In view of the amount of DDT we still gloriously sell to the African Third World, I can only imagine the horrific damage in these species' winter ranges. In addition to all this it must be said that range contraction and expansion is a natural phenomenon not yet fully understood; but it is surely and devastatingly being complicated by our chaos and apathy.

Enough misery. Back in the half light of the heath it is too dark to see any more, and it's time to go. Rubbish. At that moment a male Nightjar wends in on its long cuckoo-like wings and wing-claps a few feet ahead before floating off across the sepia sky in a wheeling flight and evaporating into darkness. And then it begins to churr. On and off, sometimes for minutes on end, it mimics a teenage easyrider on one of those irritating 50 cc mopeds as it hawks across the heath.

Miles away the giant chrome lens of the Stone Curlew has clicked open. Its iris expands to reveal a bevy of its kin in a manic party.

A mosquito makes a suicidal raid on your carotid artery and that finally ends it for you, you leave the heath. Let's hope all the treasures of the heathland are still all there in the morning.

Identification of what is around you will greatly increase your enjoyment of Natural History. The following pages describe the species that are most likely to be encountered in a Heathland habitat. Each species is also featured in a colour plate, representative of the type of habitat in which it is most likely to be found. The combination of the picture and description should enable you to find out what is flying, standing, buzzing or growing in front of you.

Heathland Residents~ A GUIDE

LEFT **Great Marsh Grasshopper**
RIGHT **New Forest heathland**

HEATHLAND RESIDENTS

EARLY-SPRING DAWN

CHRIS SHIELDS 1987

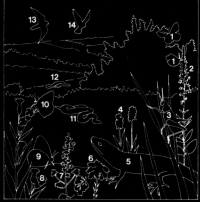

SUMMER SUN

1 Silver-studded Blue (p. 80)
2 Spiked Speedwell (p. 75)
3 Heath Grasshopper (p. 80)
4 Heath Spotted Orchid (p. 77)
5 Sand Lizard (p. 85)
6 Cross-leaved Heath (p. 74)
7 Breckland Speedwell (p. 76)
8 Sand Catchfly (p. 72)
9 Grayling (p. 80)
10 Green Hairstreak (p. 81)
11 Smooth Snake (p. 86)
12 Common Viper or Adder (p. 86)
13 Hobby (p. 87)
14 Kestrel

DAMP AND BOGGY

THE HEATHLAND UNDERWORLD

PLANTS

Sphagnum mosses

For a qualified botanist this group of mosses are readily separated from all of the other types of moss by the unique characters of their reproductive apparatus. The ripe, ovoid capsule is held on a short stalk. When ready the spores are explosively released. To the keen amateur naturalist the separation of the 13 or so species which occur in Britain, is really a matter for examination of this capsule under a microsope. To the casual explorer this genus of mosses is still easily identified by their cushion type of growth pattern which can dominate areas of wet heathland. The ground colour of the moss ranges through yellow to green and brown and even a rich red depending on the exposure and acidity of the water, and many species vary dramatically in the colour and form of their tissues under different conditions of humidity, acidity and exposure.

Marsh Clubmoss

Lycopodium inundatum Prostrate

This low perennial is similar to, but more robust than, the true mosses. It has small pointed leaves which overlap along the stems. The spore capsules, which are held in erect cylindrical cones, are carried at the base of the scales. In Marsh Clubmoss these cones are solitary and held at the top of the leafy stalks. This species is widespread but only locally abundant on small areas of wet heath in the lowlands where it grows on open, bare, moist peat and 'flowers' between June and September. During the winter it dies back to leave only the terminal bud visible on the surface.

WINTER WILDERNESS

TOP LEFT **Short-eared Owl (p. 89)**
BOTTOM LEFT **Great Grey Shrike (p. 91)**
TOP RIGHT **Merlin (p. 88)**
MIDDLE & BOTTOM RIGHT **Hen Harrier (p. 87)**

Stagshorn Clubmoss

Lycopodium clavatum Prostrate to 8 cm

This species has a creeping mainstream which sometimes extends amongst the heather and sphagnum. Its leaves are green, closely set and are toothed with white bristle points. Its cones appear in pairs on erect, long stalks and are covered in scattered yellow toothed scales. It is more common in the north and west but occurs locally in heathy woods in the south and east of England, and its 'flowers' appears between June and September.

Fir Clubmoss

Urostachys selago Prostrate to 12 cm

This species is a stout, bushy and tufted plant with erect stems which fork two or three times. The leaves are not toothed and the spore sapsules are held at the base of these whilst vegetative buds occur often near the top of the stem. Today, this species is more frequently found on the hills and mountains in the north and west and in Ireland and is largely absent from the lowlands of the south and east. Its 'flowers' appear between June and August.

Bracken *Pteridium aquilinum* 2 m

This well-known, stout, strong, tall and gregarious fern sometimes grows in dense swards under oak trees or on heathlands. Its leaves are three-pinnate with the long narrow leaflets pinnately lobed and held on a stout, tough stalk which arises from the far creeping and virtually indestructable rhizome root stock. The shoots or 'fiddle heads' appear in May and grow incredibly quickly until the Autumn when the leaves collapse to leave a copper-brown carpet of vermiculate chaos throughout the winter. Bracken prefers dry, light, acid soils and is at its peak between July and August and is wind-spread all over Britain. Comparatively recently, due mainly to the effects of burning, it has increased in abundance on heathlands where it soon shades out and destroys the native heath habitat. Thus in many areas where specific species, such as Sand Lizard, or plant species need to be conserved it is now sprayed with an effective herbicide which over a period of two or three years can reduce the bracken cover by 80–90%. Nevertheless it is still a serious pest of heathland.

Wavey Hair-grass
Deschampsia flexuosa 25–50 cm
For the amateur botanist, identifying grasses is a daunting prospect because many species can only be separated from another by minute details of the anatomy of their leaves or flowers. However, Purple Moor-grass, Sheep's Fescue and Wavey Hair-grass all have distinct qualities which can be seen at the macro level. Wavey Hair-grass flowering in June and July forms a beautiful spangled sward like an area covered with a mist of static, shining, silver sparks. Each spark is formed from a two-flowered spikelet held on a thin hair-like stalk. Each flowerhead is branched and spreading and has many of these small flower stalks, and this gives the whole grass a graceful appearance. It grows in tufts with rather short, dark green, bristle-like leaves, which have rough sheaths and sometimes, particularly on the edge of forested rides, the heathland can be carpeted with this species in a single sward which will tickle your legs as you march through it.

Purple Moor-grass
Molinia caerulea 30–80 cm
This is a coarse and variable hairless perennial which forms dense tussocks with wirey stems. Its leaves are stiff, flat and greyish and their sheaths are a deep purple. Its flowerheads appear between July and September, and are narrow and spike-like having very small purple spikelets, each holding between 1 and 4 flowers. This grass can be found all over Britain and is often locally abundant on the wetter parts of heaths and moors. In winter this grass becomes a rich amber colour and in the cold sunsets over the heath glows with

a beautiful richness, often forming dense ochre swards which snake up the heathland valleys.

Sheep's Fescue *Festuca ovina* 8–25 cm
This grass is very common and also very variable in appearance. Its hairless and perennial leaves are short, very narrow, inrolled and are almost always a waxey green. In some places, especially dryer grassland on chalk or limestone it can form a single species sward, and is renowned for being very slippery due to the waxyness of its leaves. Its flowerheads are also characteristic, even though they vary between branched, spreading and compact, because they have purplish spikelets each of which has many flowers. These appear between May and July. It grows on heathlands, as well as downland, all over Britain.

Sheep's Sorrel
Rumex acetosella 7–25 cm
This low and slender perennial is common throughout the whole of the British Isles where it is often a characteristic plant of grassy heath. It enjoys acid, sandy and well drained soils where it forms an association with Common Bent grass *Agrostic capillans* and Heath Bedstraw *Galium saxatile*. It has separate plants for each sex and flowers between May and September. These tiny flowers are held on a leafless spike, are well spaced and vary between yellow and pink depending on their age. Its leaves arise basaly and are an extraordinary lobed arrowhead in shape.

Sand or Striated Catchfly
Silene conica 5–23 cm
Occurs only in breckland in Britain. A neat, grey, stocky, downy annual with narrow leaves and a strongly-veined, inflated calyx which gives it its nominal striations. Flowers appear from May to August and vary between a rich and washy pink on small notched petals. It is found on stabilised dunes, open sandy grasslands, tracksides, and seems to favour areas with

man-made disturbances. A similar species also found in breckland refuges is the Small-flowered Catchfly *Silene galliea*.

Spanish or Breckland Catchfly
Silene otites 25–45 cm
A widespread grassland plant of central, southern and eastern Europe, reaching the north-western limit of its distribution in the English brecklands. It likes open areas of short turf where it needs rabbit scratchings or human disturbance to enable germination. A sticky, hairy and unbranched perennial; with leaves that are dark green and arranged in rosettes; the basal forms broadest nearer the tip, the rest spread sparingly up the stem. The small, yellowish-green, spiky flowers appear from June to August and occur in whorls on broad spikes. Their shape is due to the ten prominent stamens or three styles on female flowers.

Gorse or Furze *Ulex europaeus* 2.5 m
This densely spiny shrub is widespread on commons in many parts of Britain where it grows on well-drained acid soils often away from heathland on coastal cliffs and commons. It flowers all year round but more enthusiastically between April and June, its rich golden yellow flowers having a deep almond scent. Its seed pods are hairy and explode loudly when ripe and its leaves are evergreen, ridged, furrowed spines. Two other forms occur in Britain: Western Gorse *Ulex gallii* which shows a western distribution on heathlands and is generally less than 1 m in height and lacks the deep furrows in its spines; and Dwarf Gorse *Ulex minor* which is restricted to south-east England westwards to Dorset. Its peak of flowering occurs after *U. europaeus* onwards to October and it is frequently found with the grass *Molinia caerulea* and Cross-leaved Heath *Erica tetralix*.

Heath Bedstraw *Galium saxatile* Prostrate
This plant is widespread over the whole of the British Isles and is often a common

Gorse

plant of our lowland heath where it enjoys the dry, well-drained, acid soils. It often forms large mats of its hairless stalks and whorls of 4 to 6 sharply pointed leaves. Its flowers, which appear between June and August, are small and white with a sickly fragrance. Its fruits are hairless and rough and occur on short stalks in clusters along the stem.

The Sundews *Drosera* spp.
The Common Sundew *Drosera rotundifolia* occurs throughout Britain on wet heaths and moorland where it grows amongst *Sphagnum* or on bare peat. It is a highly distinctive slender perennial plant with a flat basal rosette of stalked round yellowish-green leaves covered in sticky red hairs. The flowers are white, in small heads held on leafless stems and appear from June to August. Occasionally they do not open and are thus self-pollinated. The Great Sundew *D. anglica* is sturdier and taller than the Common Sundew with

Ling, the dominant heathland plant

narrow tapering leaves rather than round ones. It is found on the heaths of the New Forest and Dorset in the south but is generally more frequently found in the north-west. The Long-leaved Sundew *D. intermedia* is much more local, scattered across the British Isles and although it too is more common in the north it can be found on the heaths of Surrey, Dorset and the New Forest. This species has a lateral not terminal flower stem which arises from beneath the rosette and which is only half the length of that of the Great Sundew. It also seems able to tolerate drier microhabitats.

Ling *Calluna vulgaris* 15–50 cm
This is it. Also called Heather, it is distributed throughout the heathlands of the cool temperate zones but can also be found in pine and oak woodland and on sand dunes. It is characteristic of these poor soils and grows in areas as high as 1,000 m. It is an evergreen dwarf shrub with numerous small overlapping ridged leaves lined in two opposite rows on the numerous branches. The pale purple flowers appear from August to September and are distinctive from *Erica* spp. because the calyx is longer and the same colour as the corolla. It is pollinated by insects, especially thrips, and is not that lucky.

Cross-leaved Heath *Erica tetralix* 30 cm
The grey-green leaves of this species are arranged in whorls of four, are short and linear, rolled down at their margin, paler beneath and fringed with long glandular hairs. The drooping rose pink flowers, which appear from June to September, are held on compact umbel heads comprising a number of flowers. Pollination is by insects and results in a downy fruit. The plant frequents waterlogged and poorly aerated peats to altitudes of 750 m.

Bell Heather *Erica cinerea* 50 cm
The glossy dark green to bronze leaves are arranged in whorls of three with smaller leaves at their bases, which is one way to

distinguish it from *Calluna vulgaris*. The flowers are also different from *Calluna* as the hairless calyx is shorter than the rich crimson purple corolla. These appear from June to October and may be self-pollinated as well as using insect vectors and the seeds they produce are held in hairless capsules. The plant is common on lowland heaths across Europe, most often growing with other ericaceous shrubs.

Other ericaceous shrubs which grow in this community include Cornish Heath *Erica vagans* which is confined to a small area of ten sites in Cornwall and one in Ireland, Irish Heath *Erica erigena* which is very local being confined to one site in Ireland, Dorset Heath *Erica ciliaris* which is found at forty sites in Galloway, Cornwall, Devon and Dorset where it is sometimes abundant on wet acid heaths, St Dabeoc's Heath *Daboecia cantabrica* which grows in west Galloway and Mayo where it is common, and an alien import *Erica lusitanica* which grows naturally in Spain and Portugal and has now established itself in Dorset and on Cornish railway embankments where its 2 m height and white flowers make it distinctive.

Harebell *Campanula rotundifolia* 12-35 cm
Known as the bluebell in Scotland, this species is generally commoner in the north of England, Wales and Scotland and only occasionally appears on the heathlands of the south and west. Its flowers appear between July and September and are a pale turquoise-blue held in a loose, green truss on long, hair-like stems. These fragile stems, the small, roundish root-leaves and the linear stem leaves give the whole plant a somewhat vulnerable appearance. On heathland it is generally found along the grassy sides of the path, because it is a poor competitor amongst the heather.

Marsh Gentian
Gentiana pneumonanthe 10-40 cm
A local and rapidly declining flower in England due to drainage, land reclamation

and birch encroachment of heathlands. It favours wet areas of the heath growing on thin layers of peat, usually with Purple Moor-grass *Molinia caerulea* and Cross-leaved Heath *Erica tetralix*, where it flowers from July to October. A weak, hairless perennial, it supports striking azure blue trumpet-shaped flowers on stems with opposite, linear leaves. These flowers are streaked with green on the outside and are pollinated by bumblebees.

Bogbean *Menyanthes trifoliata* 20 cm
This plant is a distinctive far-creeping aquatic perennial which has large hairless trifoliate leaves (like giant clover) and conspicuous spikes of white flowers, pink outside with five petals fringed with white hairs. Both leaves and flowers project above the water's surface and the flowers appear from April to July. It is commonest in the north and west of Britain but occurs locally on lowland heaths where it favours still shallow margins of ponds and deep anaerobic muds.

Common Dodder *Cuscuta epithymum*
The dodders have slender para-leafless, entwining stems which usually appear annually and attach to their host with modified suckers called haustoria. The leaves are in fact reduced to tiny scales and the thin flaccid stems twine anti-clockwise around Ling *Calluna vulgaris* and Gorse *Ulex* spp. which are their usual hosts. The flowers appear between July and September in unstalked globular heads and are bell-shaped, translucent, waxy-white tinged pink with slightly protruding stamens and styles. The calyx is reddish with pointed teeth and the plant is widespread but local south of the Thames, being especially prevalent in the New Forest and Dorset.

Spiked Speedwell
Veronica spicata 10-30 cm
In Britain the Spiked Speedwell is divided into two subspecies: *V. spicata hybrida*, a larger plant which grows up to 60 cm on

Spiked Speedwell

the limestones of Wales and northern England, where it is nearly as rare as *V. spicata spicata*, which is the Breckland equivalent, now limited to only four sites. It is an unbranched, downy perennial with slightly toothed, oval leaves of which the lower are stalked. The small intense blue flowers appear from July to September in long dense terminal spikes. It prefers closely grazed, well-drained sands, so its decline is attributable to the intensive agriculturalisation and the demise of the rabbit in the Brecklands.

Breckland Speedwell
Veronica praecox 8 cm
This is a native of southern and western Europe and this too is known from only four localities in East Anglia. It grows in bare, loose sand in fallow fields and their margins, on sandy roadsides and around rabbit diggings. It is a small annual which flowers from March to May and has leaves which are often purple below and have shorter stalks than the long ones which support the dark blue flowers.

The Spring Speedwell *Veronica verna* and Fingered Speedwell *Veronica triphyllos* are now very rare and restricted to eight and three sites respectively in the English Brecklands. Both are small, downy annuals with small dark blue flowers which grow in loose disturbed sands in crop or field edges. Again agricultural reclamation is to blame for their demise and if you find one of these then you're looking for it and it requires no more description.

Bog Asphodel
Narthecium ossifragum 18 cm
Common in north-west Britain, this species occurs locally on our lowland heaths where it grows most favourably in acid peat areas with ground water movement. It is a hairless creeping perennial, having basal tufts of sword-shaped orangy-green iris- like leaves and longer stems bearing

spikes of bright yellow star-like flowers which have tufts of white hair on their six orange anthers. These flowers appear in July or August and are pollinated by a variety of insects to produce conspicuous deep orange fruit capsules.

Wild Gladiolus *Gladeolus illyricus* 30 cm
Now very rare, this bulbous perennial is found only in the New Forest where some thirty-five populations are still picked despite its legal protection. It has rich reddish-purple flowers laced down one side of the flower spike. These are never longer than 2–5 cm, have pointed petals and anthers which are shorter than their stalks and are diverging at their base. They appear in July amongst bracken on dry heathland and are beautifully unmistakable.

Bog Orchid *Hammarbya paludosa*
This orchid is now confined to western Scotland and the New Forest in Hampshire with a few other isolated localities. It was commoner in the past but drainage of its wet acid *Sphagnum* bogs has reduced its numbers. It grows in exposed situations on top of the *Sphagnum* but nevertheless is an incredibly inconspicuous orchid. It matches the colouring of the moss and is easily overlooked. Flowers erratically between July and September.

Heath Spotted Orchid
Dactylorchis maculate 15–30 cm
This orchid is very variable in form, but has narrow keeled and folded leaves, which occur on a solid stem, and are spotted with purplish black. The flowers are held in a short, broad, cylindrical spike and are pale pink, pale purple or more often white marked with small, coarse, crimson dots and splotches. These flowers closely resemble those of the Common Spotted Orchid which is usually taller and more robust, and often darker pink in colour. They open from the bottom of the spike upwards and appears from the middle of May through to the end of July. The Heath Spotted Orchid is common and wide-

Heath Spotted Orchid

spread on damp, acid heaths and moorland, often growing in quite wet areas, in small clumps or aggregations where grazing pressure is light.

Silver Birch *Detula pendula* 25 m
This deciduous native seldom grows very large and is easily recognised by its black and white, peeling, papery bark and rugged trunk. Its twigs are shining brown and hairless, often having tony worts and its leaves are a very pointed oval and toothed. These turn from a pale green in spring and summer, to a rich yellowy-orange for a few days in Autumn. Yellow catkins appear in spring, the males being longer and looser than the females. The resulting fruits appear in the form of two winged structures. At a distance these trees often appear loaded with bird-nests in winter, but in fact these clumps of tightly growing twig are a growth deformity known as 'Witches broom'. These trees are often also victim to gall infections of the trunk

A damselfly on sundew

which produce large, ugly and worty swellings which actually do little to reduce this species's eighty-year life-span.

Many other species of plant occur on heathlands ranging from the trees, such as Scots Pine *Pinus sylvestris*, to shrubs such as Broom *Cytisus scoparius* and the hideous, introduced Rhododendron *Rhododendron ponticuum* and numerous other flowers including the Tormentil *Potentilla erecta*, Pale Dog Violet *Viola lactea*, Heath Dog Violet *Viola canina*, butterworts *Lentibulariaceae*, Greater Bladderwort *Utricularia vulgaris*, the hawkweeds *Hieracium* spp. and a host of specific lichens which encrust the surface at various stages of the heath's development and can be excellent indicators of the degree of air pollution that the heath is subjected to. To identify these and other species please refer to one of the guides mentioned in the bibliography, although a more specialist text might have to be used.

INVERTEBRATES

Small Red Damselfly
Ceriagrion tenellum 3 cm
This tiny, weak and primarily red-coloured damselfly is on the wing from May to September with a peak of activity at the end of June. It occurs on the south and west heathlands where it is sometimes common. The females are more variable in colour than the males which have a black thorax, although wholly red or black forms do occur. It rarely moves far from its larval habitat which is the shallow edges of acid pools lined with *Sphagnum*.

Scarce Blue-tailed Damselfly
Ishnura pumilio 2.5 cm
Quite similar to the much more common Blue-tailed Damselfly *Ishnura elegans*, but males can be told from these by a thin black line bisecting their blue tails, which are a small powder blue spot at the end of a black abdomen. The thorax is black and

blue striped and a tiny blue eye spot is present. Females are green where the males are blue and some segments of their abdomen are reddish-bronze. This species can be seen about the bog pools and slow streams of Hampshire, Dorset and Cornish heaths between May and October.

Southern Damselfly
Coenagrion mercuriale 2.5 cm
Slightly larger than *Ishnura pumilio*, this species has pronounced thoracic stripes and two pronounced eye spots which are joined by a tiny blue line. The abdomen is banded with the same pale powder blue and black and the females appear almost entirely black from above. With closer examination they have green where the male is blue. This species is on the wing from May to August over the valley bogs of the new Forest where it is common. Elsewhere in the south it occurs sporadically, flying slowly amongst low vegetation and over its favoured bare wet areas of heath flushes.

Keeled Skimmer
Orthetrum coerulescens 4.5 cm
This species has a pronounced affinity for *Sphagnum*-filled ponds on heathlands as long as some open water is available. They fly rapidly and erratically along the edges of the ponds where they are not territorial, sometimes occurring *en masse*. Their wings are held forward and downward at rest and in males the slender abdomen is a bright blue, and the thorax dark brown. Females are almost entirely yellow-brown and in both sexes the wings appear very long, especially in flight. It is commonest in the south especially in the New Forest where it flies from June to September.

Black Darter *Sympetrum danae* 4 cm
This is primarily an upland species in Britain but it can still be found locally between June and October on some of our lowland heaths. Colouring varies according to age, the males being yellow and black to begin with, becoming all black

when mature. Females are yellowy-brown marked with black and often a large accumulation of mixed sexes will occur over their favoured *Sphagnum*-filled bogs, streams and flushes. These can be closely approached and if they take flight will soon land again.

White-faced Darter
Leucorrhinia dubia 4.5 cm
A rare British species, it occurs on some Surrey heaths and also in northern England and Scotland. Both sexes have a white face. The males have a body marked with scarlet and in the females this red is replaced with a pale creamy-yellow. It occurs over shallow peaty pools in heathland but its habit of flying low over the heather and always settling makes it inconspicuous from late May to late July.

Note that the damselflies and dragonflies described here are the more specialised forms often restricted to heathland and are consequently rare. To identify the more frequently encountered, cosmopolitan and common species reference to a specific guide is required (see bibliography).

Large Marsh Grasshopper
Stethophyma grossum 3 cm
This species is one of our rarest grasshoppers but can be found in the *Sphagnum* filled bogs of parts of Dorset and the New Forest, East Anglia and several other southern counties. It is difficult to find because it lives in the wettest part of the bog, and flies readily on hot days. Also, once located on the *Sphagnum*, its camouflage is quite cryptic since its body is marked with the same range of yellow-green and reddy-brown of that of the *Sphagnum*. It stridulates quite loudly producing a pronounced ticking sound. Usually about 8 ticks are produced at a rate of 1–3 per second. It can be found most often in August and September and is the largest of all the British grasshoppers. Only ever really active on bright sunny days, care must be taken pursuing these insects over

the 'quaking' bogs they favour when Wellingtons are useless and a hindrance.

Heath Grasshopper
Chorthippus vagans 1.4–1.8 cm
This rare species was only first identified in Dorset in 1933. Since that time it has been found on a few more sites in this county and also in the New Forest (Hampshire). It is a medium-sized grasshopper which is dark grey or grey-brown in colour and it often has slight mottled forewings. In mature specimens the upper part of the abdomen is tinged with an orange-red flush and this spreads to the tibia part of the legs. It favours dry sandy heaths dominated by Lings and appears in August, occasionally persisting through to October. It sometimes cohabits with the Mottled Grasshopper with which it is easily confused unless the sex is known. Stridulation is in a burst of 3–8 seconds of uniform loudness and the sound closely resembles that of the Meadow Grasshopper. One for the experts!

Gorse Shield Bug
Piezodorus lituratus 1–2 cm
This species is frequently found on Gorse, Broom or Dyer's Greenweed, particularly in the south of England. It belongs to the Pentatomidae family which is on the edge of its range in Britain. New generation adults occur from late July onwards. Initially they are pink and green but after overwintering they become bright green with the onset of sexual maturity. Their courtship is complex and the males stridulate like a grasshopper. Eggs are laid in two rows of seven each on gorse spikes and the incredibly cryptic larvae eat the Gorse seed pods. These species enjoy hot summers and Gorse Shield Bugs become much more numerous after a run of prolonged dry summers; our wet winters also make them prone to fungal attack. These insects are separated from other bugs with shield shape bodies by their green triangular scutellum which tapers to a point midway down their backs.

Heath Assassin Bug
Coranus subapterus 0.7–1 cm
These insects are voracious predators which feed on other arthropods, and are thus equipped with a sharp stylet which can give a handling human a painful needle-like jab. Most of the family members are tropical, with only six species occurring in Britain, of which the Heath Assassin Bug is the most common. It is a blackish-brown insect, with a clearly segmented abdomen and small vestigial wings which lay like tiny glass-plates on its otherwise hairy back. Its legs are black, subtly banded with a paler brown and it is conspicuously fast moving. They also strigulate loudly like grasshoppers and appear from July until October on the edges of heathland paths. Mating occurs after a summer of feasting and the species overwinters as eggs which lie in the moss litter layer.

Grayling
Hipparchia semele Wingspan 5 cm
This butterfly is widely distributed in Britain on heaths, commons, rocky hillsides and also by the seaside on rocky coasts. The caterpillar is pale yellowish-brown with darker stripes and feeds on hair grass *Aira* spp. and a range of other grasses. The adults are on the wing from July to September and have a speedy flight, frequently landing on paths or other bare areas where they vanish due to their cryptic colours and behaviour. The undersides of the forewings are pale straw coloured with two black and white eye spots, and the hind-wings are a vermiculate mass of black and brown. Above the general colouring is a light bronzy-brown marked with cream, orange, dark brown and black, but you will never see these patterns naturally since it always snaps its wings closed on landing.

Silver-studded Blue
Plebejus argus Wingspan 2.5 cm
The sexes are distinct. The males are light violet blue, both fore- and hind-wings

edged with black and white fringes and have a row of black spots on the hind-wing. Females are bronzy-brown, be-spotted with brown and orange and the undersides of both sexes are marked with spots over a pale coffee-coloured base. As adults they live for about two weeks between July and August, have a weak flight and enjoy periods of basking in the sun. The caterpillars are olive-green with reddish stripes and feed mostly on gorse *Ulex* sp. and this species is sometimes locally very abundant on its favoured haunts of sandy heaths.

Green Hairstreak
Callophrys rubi Wingspan 2.5 cm
This is our commonest and most wide-spread hairstreak butterfly which can be found on scrubby areas over most of Brit-ain. It occurs on heathland because its yel-low and green striped caterpillar feeds on the developing fruits of gorse *Ulex* sp., among other things. The adults are on the wing from May to July flying rapidly over short distances. They are very difficult to locate when at rest due to their green under-wings. Their upper-wings are rich bronzy brown and they can be seen twist-ing up from their perches to perform pretty aerobatics before returning to sunbathe.

Fox Moth
Macrothylacia rubi Wingspan 5–6.5 cm
The Fox Moth, one of the Eggar family Lasiocampidae, is widespread on heath and moorland. The male flies both day and night, whilst the female only flies at night, from May to June. The adults are a fawny beige, the male being a little darker brown, with two thin creamy stripes on the forewings. The caterpillar is a large black form with a thick covering of rich brown hairs and a line of more reddish hairs down the back, and is common on heathlands. The hairs are a highly effective irritant. The caterpillar is frequently parasitised by Braconid Wasps, whose larvae can be seen emerging from its body to spin their little cocoons on its sides.

Emperor Moth
Saturnia pavonia Wingspan 5.5–6.5 cm
Our only representative of the silk moth family. In Britain this species is wide-spread and very common on heathlands where adults are on the wing from April to July. Both sexes have large dark eye spots on both the upper and lower sides of the wings, larger females being generally browny-greyish with bands of cream run-ning down the wing edge. The males, with their conspicuously fluffy antennae, are browner and have rusty underwings. The caterpillars are initially black and hairy, maturing into green and black spotted forms with hairy yellow warts, feeding mainly on Heather *Calluna vulgaris*. The females fly only at night but males can be seen on the wing during the day.

Ruby-tailed Wasp *Chrysis ignita* 1 cm
One of the Chrysididae family of parasites whose hosts are various solitary bees and wasps. This species is brilliantly coloured, with a rich metallic emerald head and tho-rax and a ruby abdomen which provides the derivation for its common name. It is common on English heathlands but not frequently encountered without specific searching. In Europe the all vivid green *Stilbum cyanurum* has a widespread dis-tribution and similar parasitic habits.

Turf Ant
Tetramorium caespitum 0.2–0.3 cm
There are thirty-six species of ant in Britain, and the Turf Ant is often the domi-nant species on the heathland along the south coast of England. It is a continental species on the edge of its range in Britain and differs from most`ants, which collect insects and scavenge, by collecting and storing seeds in underground galleries. It also defends territories of up to 80 sq metres and lives in deep nest which enables it to survive hot, heathland fires and dessication during the warm summer months. This species is frequently a victim of the slave making ant, *Anergetes atratulus*, which has no workers of its own

at all and survives entirely parasitically. The *Anergetes* queen kills the *Tetramorium* queen and uses all her workers to raise her own sexual generations. These mate in the host's nest before the hundreds of fertilised queens fly off to other nests to continue the piracy. After the Turf Ant queen has been killed there is no recruitment to this species' worker population because no eggs are produced, the colony gradually declines and after two years it dies.

Potter Wasp
Eumenes coarctatus 15 mm
The sole representative of the second of two genera of solitary vespid (true) wasps in Britain, this species is in no way colonial although the females do construct neat little mud capsules for their young and this activity provides them with their common name (see page 00 for precise description of this part of its biology). It is a black and yellow wasp with a banded abdomen not unlike the common wasp in pattern. It can be found with a valiant amount of searching on southern heathlands, where it leaves its handiwork on the fronds of *Calluna*. It has a group of nineteen British cousins, the *Odynerus* or mason wasps, and a larger European counterpart in *E. unguiculata* which is nearly twice as big, lacks the black markings being primarily a tan brown and ochre.

Sand Wasp *Ammophila sabulosa* 2.5 cm
Otherwise known as 'digger' wasps because of their excavating habits, this species is one of the British representatives of the Sphecidae family, all of which are typified by the shape of their abdomens, extremely narrow in the front and ending in a bulbous knot. In these British species the middle part is always orange and the rest of the body is a shiny black. They are easily differentiated from other insects by their rapid dizzy flight and manic attitude when running over the sand, and from other hunting wasps such as the spider hunting wasps Pompilidae, of which there are about forty British representatives mostly favouring sandy areas, by their abdominal structure and reference to a more precise guide (see bibliography). Locally abundant on southern heaths, *Ammophila* seems to have favoured haunts, probably untrampled sands where their burrows remain undamaged. They are also strongly sexually dimorphic.

Heather Beetle *Lochmaea suturalis* 5 mm
The most important phytophagous beetle on the heath, this species has long been known for its habit of excessive, intense defoliation of heather over large areas. On the northern moors this is an annual problem, but it is a much less frequent phenomenon in the south. Adults are on the wing in March and April and their offspring feed on the *Calluna* shoots in the summer before they pupate to re-emerge for another adult feeding frenzy from September to the winter.

Green Tiger Beetle
Cicindela campestris 15 mm
This beetle is by far the commonest of the five species of tiger beetle which can be found in Britain. They differ from members of the Carabidae by having no obvious striation of their elytra or backplates. All of these species favour open sandy or gravelly paths on which they bask until you disturb them. They then rise in a rapid buzzing flight only to return to the path a few metres further on. When you come to investigate they will make good their escape by sprinting away on long legs. The Green Tiger Beetle is very variable in colour but typically is a brilliant burnished emerald with coppery legs and underside. It can be found on many heathlands in southern England, becoming scarcer further north.

Ladybird Spider *Eresus niger*
Male 8–10 mm, female 9–15 mm
Impossible to see in Britain, this species may be encountered on holidays in southern Europe if you believe in miracles. The

A water boatman sticking to the water surface

males are unmistakable with their white-banded black legs and their brilliant orange-marked abdomen with three pairs of black spots. These males are very short-lived as adults and the females rarely leave their hiding places which are inconspicuous tubular burrows. If they do emerge they are dark grey with a dusting of lighter grey pubescence. They prefer south-facing slopes, well protected from the wind with shrubs and stones.

Pink Crab Spider *Thomisus onustus*
Male 3 mm, female 8 mm

If you can bear to search, this species can be found on the southern English heaths and over most of Europe. The females are both variable and changeable in colour from pink to white, but usually a combination of these depending on the plant they most often rest on. On our heaths the Cross-leaved Heath *Erica tetralix* seems a favourite. Her abdomen is a triangular prism and her eyes are raised on tubercles behind which a pale stripe usually runs centrally down her carapace. Males are warty dark brown, apparently 'starved' similes of their mates. June to September.

Raft Spider *Dolomedes fimbriatus*
Male 13 mm, female 22 mm

Large, sturdy and aquatic spiders which are fairly common locally in swamps and heathland mires in southern England. In the north of England they occur less frequently, whilst in Europe they are widespread. The body and legs are primarily brown but it has a pair of distinctive pale cream bands running from the eyes to the rear of the abdomen, the males being more clearly marked than the females. Females can lay up to 1,000 eggs, so at times the young occur in masses, often away from the water in adjacent vegetation. *D. plantarius* is a similar species which is a little more variable colourifically but as it is very rare in Britain, occurring only in Suffolk and Norfolk, confusion is unlikely. On the continent it is widespread but not as common as *D. fimbriatus*.

Pisaura mirabilis
Female 12–15 cm; male 10–12 mm

These spiders are readily found on the grassy parts of heaths where the females construct a conspicuous tent-like web which acts as a nursery for the young spiders which emerge from her carefully guarded egg sack. Both sexes are generally a grey-brown, paler on their sides, and often marked by a darker central stripe on the tapered abdomen and carapace. This is edged with lines of pale cream, and often the eyes have conspicuous tear marks at their edges. The spiders are long-legged for rapid movement through their grassy habitats and are the only member of this genus in northern Europe. They breed in early June and by the end of the month and during July the females can be seen carrying large cream egg sacks in their chelicerae. Interestingly the males are the only British spiders to present the female with a courtship gift. This species is very common and widespread all over Britain and in Europe in dry, grassy environments.

Other notable heathland invertebrates include three thrip species which feed on heather pollen, and the Bog Bush Cricket *Metrioptera brachyptera*, the most common heathland bush cricket.

Other butterflies include the Meadow Brown *Maniola jurtina*, the Gatekeeper *Pyronia tithonus*, the Common Blue *Polyommatus icarus* and the Small Heath *Coenonympha pamphilus*, and other moths include True Lovers' Knot *Lychophotia porphyrea*, Beautiful Yellow Underwing *Anarta myrtilli*, Narrow-winged Pug *Eupithecia nanata*, Ling Pug *E. goossensiata* and the Common Heath *Ematurga atomaria*.

About twenty-five species of carabid beetles join the others, the Violet Ground Beetle *Carabus violaceus* being the most frequently encountered, as are Oil Beetles *Meloe proscarabaeus* and Minotaur Beetles *Ceratophytes typhaeus*.

A rich, complex community of Hymenoptera (bees, wasps and ants) is typical of most heathlands and, apart from those already described, include the spider-hunting wasps Pompilidae, a bevy of bumblebees *Bombus* spp. and the handsome Velvet Ant *Muttilla europaea*, which despite its name is in fact a type of wasp, and numerous species of real ants. Hoverflies Syriphidae are common around the blooms of gorse and heather and the wonderful and huge Horsefly *Tabanus sudeticus* occurs in the New Forest.

Many other species of insect occur on heathland, and immigrants from surrounding habitats can also occur, adding to the numbers. For precise identification it is probably best to consult one of the more specific field guides mentioned in the Bibliography, but even these contain only a representative sample of species that you may encounter.

VERTEBRATES

Palmate Newt *Triturus helveticus* 9 cm

This newt can often be found in heathland pools and puddles and even in the deep ruts formed by tractors during the process of afforestation. They are usually smaller than their quoted size and the males are the smallest of the sexes. They are a smooth-skinned newt with three grooves visible on their head, and are olive or pale brownish above, often having two lines of small spots along their back, and a dark stripe on the sides of their head. The colour on their underside is restricted to a single yellow or silvery orange stripe which runs along their belly. In breeding conditions the males have a very low, smooth-edged crest on their body and tail which ends in a single dark filament. Their hindfeet also become dark, strongly webbed and their tails acquire a central band of orange which is bordered by two rows of large spots. They could be confused in England with Smooth Newts, but have unspotted throats of translucent pinkish flesh, or in Europe with Bosca's Newt, but this species has a much brighter belly colour

A male Sand Lizard

and a well spotted throat. The Great Crested Newt *Triturus cristatus* which can grow up to 14 centimetres in length also occurs in many of the southern heathland pools in Britain where it was once thought to be quite rare. It is however locally common and easily distinguished from the Palmate Newt by its huge size, dark, coarse skin, and a yellow to reddy-orange blotched belly. The males have a high, spikey crest and develop a bluish streak along their tails in the breeding season.

Natterjack Toad *Bufo calamita* 8 cm
This sturdy, short-limbed toad has prominent and parallel paratiod glands, a golden eye with a horizontal pupil and a distinctive yellow-cream stripe which runs down its back. Males have an external vocal sac which they use just before sunset to make a loud careering rattle which lasts for 1 or 2 seconds both beginning and ending abruptly. It is a runner, not a jumper, and is found in areas of warm loose sand and on heathlands south from Scandinavia to Spain, where unusually large individuals often occur. In Britain it is now very restricted, occurring at only two indigenous inland sites and a scattering of coastal localities on the east and northwest coasts of England.

Sand Lizard
Lacerta agilis 22.5 cm (Torso 9 cm)
This species is easily distinguished from the other European green lizards and in Britain from the only other lizard, *Lacerta vivipara*. It is a short-legged, stocky lizard with a short deep head (especially noticeable in males). The colouring is extremely variable but usually there is a light central streak to the dark central blotchy band

running down the back, either side of which the males have green or yellowish-green flanks which are more intensely coloured when breeding. These sides have a spattering of spots (ocelli) and mottlings. Females have grey or brown dark central bands which may be broken up and at times these too may have green flanks, though they are more usually brownish. The underparts are whitish, greenish or yellow bespecked with black, more distinctively in males. The young are more dilute colourifically than the adults, totally lacking in green.

They can be found in southern England especially in Dorset and Hampshire and a colony still survives in the north-west of England. In Europe they can be found on heaths south from southern Scandinavia, through Denmark, Germany, the Netherlands, Belgium and France, but not into Brittany or down the Atlantic coast or in northern Spain, often occupying a variety of other dry habitats ranging from road verges to crops.

Slow Worm *Anguis fragilis* 50 cm
This legless lizard is one of only two European species from a family which is predominantly confined to the Americas. The only other Anguidae, the European Glass Lizard *Ophisaurus apodus*, occurring in Europe is only found in the Balkans and parts of south west and central Asia. Consequently, Slow Worms are unmistakable, smooth-scaled, snake-like reptiles which are usually brown or grey, or even metallic copper above. Females have a ventral stripe and rather dark sides and belly, and some males occasionally have blue spots. The young are striking little golden needles, heavily marked with black on their sides and belly and although some individuals can grow up to 50 cm most are usually smaller. Slow worms are often encountered on heathlands where there is plenty of grass and are most readily located beneath sun warmed objects, such as sheets of old iron or stone. They are non-poisonous, but if handled care

should be taken since their tails are often dropped in a vain attempt to distract the human predator.

Smooth Snake *Coronella austriaca* 60 cm
Confusion should only really occur with the Southern Smooth Snake *Coronella girondica* but this could only occur where their ranges overlap in northern Spain and south-western France. Its lack of keeled scales and vertical pupil combined with an altogether different neckless smooth body form make it clearly distinctive from the Adder *Vipera berus*.

This is a relatively small snake with a cylindrical body, which appears almost neckless, and a small pointed head. It has small eyes with round pupils, smooth scales and variable colouring. It is generally greyish, brownish, pinkish or even reddish, overmarked with small dark spots and mottlings which often fuse around the neck to give a collar or blend into transverse bars across the back. The belly is darker. It is a slow-moving, slovenly animal which is apparently intelligent, as snakes go, and it frequently bites. It constricts its prey, which is primarily other reptiles and small mammal nestlings. Confined to the central southern heaths in England, it is found on embankments, in open woods and their edges and dry bushy slopes in its continental range where it occurs from the heaths of northern Spain to those of southern Scandinavia.

Adder or Common Viper
Vipera berus 65 cm
Very distinctive with its clearly marked dark zig-zag vertebral stripe it could really only be confused with Orsini's Viper *Vipera ursinii* which has a narrower head with distinctive scaling, and smaller adult size and never occurs on heathland, and the Asp Viper *Vipera aspis* which does occur on heathlands in France and northern Spain. This species generally has a pronounced upturned snout, again negating improper identification. The colouring of Adders varies according to sex; males

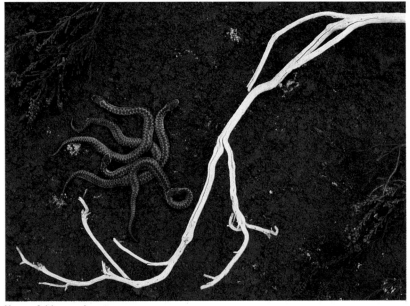

Young Adders on burnt gorse

tend to be more contrasting, often silvery-whitish with black markings, whilst the females tend to be more earthy, being found in a range of browns. They mainly feed on small mammals and lizards, swim well, and are found on all the northern European heaths except for those in Ireland and western France. It is, of course, venomous and a few human fatalities have occurred in exceptional circumstances. Nevertheless they are easily observed from close range on warm summer mornings before it is too hot and they retreat to the shade. Care should be taken at all times, but these animals are not dangerous unless you are.

Hen Harrier *Circus cyaneus*
43–51 cm (Wingspan 100–120 cm)
This species is strongly sexually dimorphic: the males are pale ash grey above with black wing tips and a white rump, the females are darkish brown with conspicuously barred tails and the same prominent white rumps. Flight is buoyant on long wings with five or six flaps and then a wavering glide, the wings being held in a constant shallow 'V' angle or flat when gliding. They can be distinguished from all other harriers by their five, and not four, apparent primaries (discernible as fingers projecting from the ends of the wings) and otherwise from Montagu's Harrier by the male having a white rump and unbarred wings and belly, and females and juveniles by their heavier build, broader wings and tail, and more definitely streaked underparts. A winter visitor to heathland, they are generally silent and may colonially roost after spending short days quartering the heath for mammals and small birds. These are regular visitors to most British lowland heaths between October and March, especially to those which are on or near the coast.

Hobby *Falco subbuteo* 30–36 cm
A small Kestrel-sized falcon with long,

A heathland winter visitor, the Merlin

sharp, scythe- like wings and a very short tail. If visible, it has characteristic and conspicuous russet undertail coverts, a generally streaked appearance below, a pale chin and a conspicuous dark and pointed moustache. It is a uniform slate grey above. Juveniles are a dark blackish-brown with heavier streaking below and lack the russet. At rest it is compact and slender but in flight, when it is most readily identified, it is dashing, elegant, supple and even mercurial in appearance. It looks like a huge swift and flies with stiff but regular wing-beats dispersed between periods of low gliding. It will soar on angled wings in thermals, unlike Kestrels which are straight-winged, but it is more frequently sighted as it 'hawks' for insects over the wetter areas of heath. It shows regularity of feeding, appearing at the same place at the same time in search of its prey, which is chiefly birds with a supplement of insects which it eats on the wing. It is migratory and arrives in northern Europe from Africa in early April and remains until September while it uses an old crows' nest to lay its 4–5 eggs. Very secretive until the young are advanced, it is generally silent away from the nest. Exciting food passes are performed at this time, but on no account should the birds be disturbed since they are decreasing in many parts of their range, and are specially protected. In Britain the heaths of the New Forest will provide your best chance of observing Hobbies.

Merlin *Falco columbarius* 27–33 cm
The size of a Mistle Thrush, the male Merlin is Europe's smallest bird of prey. Its short square-cut tail, broad-based short pointed wings and blue-grey upper colouring make it distinctive from the other falcons. If confusion is possible, it is most likely to occur with the Sparrowhawk, since its flight is bold and dashing on fast wing-beats and short glides. When hunting its flight is straight and purposeful and Merlins perform exciting and erratic aerobatics when chasing their prey, which is

primarily pipits, buntings and finches. The female is deep brown above, similarly compact with a heavily barred tail and heavily striped creamy underparts.

These falcons seem to like perching on low vantage points and are most readily located as fast-moving brown things annoying finch flocks in autumn and winter white skies. Rapidly decreasing in most of their breeding haunts, they seem to be often seen wintering on coastal flats nowadays, but a few certainly spend time on inland heaths.

Stone Curlew
Burhinus oedicnemus 41 cm
This bird is unmistakable. Firstly it is nothing like a Curlew. It has a short, straight bill, black at its tip and yellow at its base, is stocky and about Woodpigeon size, being supported on long pale yellow legs and it has a large flat-tipped head punctuated with a huge yellow eye. Its plumage is generally sandy with a variable, distinct black and double white wingbar. In flight its legs trail and it flies with slow deep angled wing-beats, with occasional quicker beats, and long glides. However, it seems to prefer to run furtively away if disturbed, head held low and body hunched until it breaks into a creeping gait, finally prostrating itself on the soil and relying entirely on a motionless crypsis to avoid predation. Otherwise its behaviour is plover-like, frequenting sandy heaths, dunes, chalk and flint arable downlands and shingle banks from mid-April to August. In Britain it has declined seriously and to see it visit Weeting Heath reserve in Norfolk and avoid disturbing it in any other of its remaining haunts.

Short-eared Owl *Asio flammeus* 38 cm
By far the most likely medium-sized brown owl to be seen flying in daylight. A patchy-looking buff bird, it has a pale tawny body with boldly streaked underparts and conspicuously long barred wings with dark carpel patches on their upper and lower surfaces (Long-eared Owls only have a patch on lower side of wing). Its ears are in fact so short that they are always invisible, its flight wavering on crooked wings, both low and rolling and with frequent periods of gliding on slightly raised wings. It could thus be initially confused with a female Hen Harrier and paler individuals are even superficially seagull mimics. They perch chiefly on the ground and are then told from the Long-eared Owl by their fawner colour and brilliant yellow eyes set in black sockets, giving them a fierce, demonic appearance. Although they hunt mostly at dawn and dusk they are also often flushed from the ground where they sometimes roost in small parties. Locally, they are regular visitors to inland heaths in winter, usually at low densities.

Nightjar *Caprimulgus europaeus* 27 cm
At rest the bird is an incredible cryptic device formed of a flattenedd head and generally elongated appearance designed to resemble a fragment of broken wood. In flight it appears as a silently flying, long-winged, long-tailed, cuckoo-like bird lacking the barred underparts. Its plumage is vermiculate in the extreme, a mass of brown, buff, grey and blacky-brown spots and bars, the males being distinguished in flight by their three white spots on the wing and two on the tail. The flight is buoyant on angled wings, and nightjars may hover when wing-clapping during the breeding season. Above all, though, it is their song which is most conspicuous. From late May to August go to wooded heathland edges at dusk to hear the monotonous churring note (see page 00) and if in southern Europe do not confuse this with the song of the mole cricket! Other flight calls are frequently heard as they hawk over young plantations and birchy areas; a soft nasal *'goo-eek'* and a high alarm *'quick, quick, quick.'* They are ground nesters, laying two marbled eggs in broken debris.

Green Woodpecker *Picus viridis* 33 cm
Despite its name the Green Woodpecker actually spends more of its time foraging out in open country, where it is particularly

fond of feeding at ant's nests. Ants are common on heathland and so as you wander across the heath these noisy and brash birds often explode a few yards in front of you and undulate away giving their characteristic, loud, ringing laugh. They are easily distinguished from the other two British woodpecker species by their larger, pale grey undersides and bright green upper parts. Adults also have a conspicuous yellow rump and lower back, and a crimson crown, while the juveniles are paler, distinctly spotted and barred. This species can be seen hopping heavily in an upright position, particularly on open, grazed lawns that occur in many parts of the New Forest and other southern heaths, as it searches for ants' nests.

Woodlark *Lullula arborea* 15 cm
Smaller, dumpier and with jerkier movements than the Skylark, the Woodlark lacks its cogener's white outer tail feathers and has a shorter tail and less broad and stubby wings. Its bold crest is often raised and dropped and it has conspicuous white eye stripes which meet on its nape. It also has a small white bar on its carpel (leading edge of wing), which is lacking in the Skylark, and has a finer bill. It is most easily located and identified in March when it is found in pairs on areas of short heather, particularly recently burned heath, foraging in a mouselike manner for its insectivorous diet. Its song is most characteristic, delivered as it soars in wide spirals and finally plunges to the ground or from overhead wires and gorse; it is a very sweet descending yodel of scales '*lu, lu, lu, lu . . . lu*'. This can be heard after dark, and Woodlarks also have a flute-like '*toolooeet*' flight call. Territories are occupied from late February and the nest is notoriously well hidden in the base of a gorse or heather bush. The hen incubates 3–4 eggs for two weeks, leaving early at dawn and dusk to feed with her mate. Strongholds for this species are now the Suffolk heaths and the New Forest in Hampshire but the species is sensitive and uncommon.

Meadow Pipit *Anthus pratensis* 14.5 cm
Tree Pipit *Anthus trivialis* 15 cm
Superficially so similar, in reality these two species are in fact readily separable in the field. They are both small, slim, long-tailed insectivorous ground nesters who are both used as a host by the cuckoo, although the Tree much less than the Meadow. Both are dusty tawny brownish but they do show the following distinct differences:

1. The Meadow has dark, brownish legs compared with the Tree's light pinkish legs.
2. The Meadow has a longer rear toenail; not a very good field identification point!
3. The Tree has a more distinct 'necklace' wing bar.
4. The Meadow has a more olive upperside and whiter, less yellow breast with smaller and more numerous streaks.
5. The Tree is plumper, tends to perch in trees more frequently and has a stouter bill.

Frankly some of these features are useless unless you have two dead pipits in your lap. For me the most useful summer guide in the field is:

6. The way they finish their display fights. Both rise into the air, the Meadow from the ground and the Tree from a perch uttering a rapid succession of notes with little melody. To complete the display they parachute downwards and before landing they make clearly different calls. The Meadow Pipit ends with a musical trill and the Tree Pipit with a '*seea . . . seea . . . seea . . .*' call.

Both birds are regularly flushed on almost all the British heaths, rising erratically, sometimes in groups, see-sawing to gain height only to fall back to the ground.

Wren *Troglodytes troglodytes* 9–10 cm
These familiar cheeky and noisy little insectivores are often the commonest

birds on the heaths. They frequent gorse clumps where they find a readily available supply of insects throughout the year and from here they often emit their loud, harsh churring when alarmed. From a distance they are easily identified by their over active foraging technique as they ricochet amongst the litter and dense canopy of the gorse, heather or bracken. They also build their bulky globular nest in gorse, generally from grass, mosses and lichen, gathered on the heath. After a particularly severe winter they can disappear from vast tracts of heathland, although they soon recolonise after a couple of warm, wet summers.

Great Grey Shrike
Lanius excubitor 24 cm

This, the largest European shrike, is a winter visitor to lowland heaths where it finds its diet of small birds, insects and occasional small mammals. The sexes are alike, and very monochrome, having a grey forehead, compared to the black of the Lesser Grey Shrike, grey crown, short broad and rounded wings with variable white panels showing as two white wingbars at rest, and a long and graduated tail. From very close range females have very faint wavy bands on a plummy chest and both sexes have a larger, more slender bill and more white on their scapulars than the Lesser Grey Shrike. Flight is feeble and the wings are frequently flicked as they undulate over large distances to the conspicuous and regular perching sites. Here they sit upright flicking their tails, scheming about impaling their prey and occasionally rattling like Magpies. They are regular but erratic visitors to eastern and southern heathlands between November and March.

Dartford Warbler *Sylvia undata* 12.5 cm

This dark grey skulker is told by its tail. Held high in flight and constantly fanned and cocked when perched, it is a very long graduated white-edged affair which distinguishes the species from any other European warbler. The flight is weak, whirring and bounding on stubby wings, generally low over heather and straight into gorse from where the bird utters its characteristic scolding metallic *'tchur-r'*, short *'tuc-tuc'* and rattling alarm calls. The song, sometimes given from more exposed perches or in a dancing flight, is a short musical chatter and is a little more pleasing than the Whitethroat's utterings. Usually seen plummeting into cover, it is best to stand off and wait for the birds to emerge revealing their often raised crests, white-spotted chins, red eyes and purplish-brown underparts. It is most obvious when feeding fledged young. On no account should it be disturbed as it is extremely rare in England, occurring only in Dorset and the New Forest.

Stonechat *Saxicola torquata* 12.5 cm

This noisy little bird perches prominently on most heaths in summer atop gorses and pines, constantly flicking its tail and flinching its wings in an upright stance. Flight is low, whirring and jerky and it is told from the Whinchat by its tiny size, dumpy profile, short wings and all dark tail which lacks any white feathers. The male is very obvious with his black head, white collar and chestnut chest, while the female is streaked brownish above and rich orange below and lacks the Whinchat's white eyestripe. Both sexes show a white wingbar in flight whilst the juveniles are more streaky versions of the female.

Their call is a harsh, grating and scolding *'tschack tschack tschack'* sounding like two stones knocking together (hence the name), and the song is a husky scratching string of irregular, rapidly repeated, double notes, not unlike Dunnock or Whitethroat, which is often delivered in a short dancing flight. A ground nester from late March to July, it lays 5–6 pale blue eggs which are incubated by the female. Showing a more westerly distribution in Britain, where the coasts are its strongholds, it also occurs on most southern and eastern lowland heaths.

Whinchat *Saxicola rubetra* 12.5 cm
These small Robin-like thrushes have a stocky short-tailed appearance which is overall paler than their cohabiting confusers, the stonechats. Males are separated from these by their prominent white or buff eyestripes and white on tail and females by their paler appearance, pale eyestripes and smaller white wing patches. Their flight is low and jerky, with a flicking of wings and tail upon landing on bushes to sing their pleasing little warble, recalling the Redstart or Stonechat. This is a fast crackly twittering which is so variable that it is incoherent and they may thus be overlooked. These insectivores are April to October summer visitors to heathland where they must be considered uncommon, despite being numerous locally, particularly in some northern and western upland haunts. Only low densities are found on our southern lowland heaths.

Wheatear *Oenanthe oenanthe* 14.5 cm
These dapper little chaps skim low over the ground and flop into a series of pretty little jumps before landing. Constantly bowing and bobbing before scurrying to a higher vantage point, such as a boulder or dried dung, to flick and wag their tails in an erect posture, they are very active at all times of the day. Small thrush-like birds, they have a conspicuous white rump and white in the tail, the males having a grey crown, white eyestripe and black cheeks, a blue-grey back and black wings and the females being a generally drabber buff and brown with less distinct markings around the eye. Whilst there is wider variation in the adults' plumage, juveniles have rich buff on their backs and rufous on their wings.

The song is a loose warbling wheezing of notes rather like stones being rubbed together and the agitated call is a familiar hard 'clack-weet'. They must have open areas and prefer short rabbit-chewed turf, often choosing a rabbit hole to nest in. Six pale eggs are laid in late April and the male often sits over the nest and dances like a yoyo while the hen completes her two-week incubation. A migrant, it arrives back on the northern European heath in early March and leaves in late October. Wheatears have declined drastically in the south and east due to the demise of sheep farming which provided their favoured short turf. Nevertheless low-density populations occur on the Hampshire, Suffolk and Norfolk heaths.

Linnet *Carduelis cannabina* 14 cm
The male Linnet can appear a very scruffy and dishevelled bird, but when in good plumage its pinkish breast and crimson crown make him very attractive. His wings are dark brown and his back chestnut brown, whilst the underparts are buff coloured streaked with brownish-black. The tail is conspicuously forked and edged with white and the head in the breeding season is a pale lavender-grey. The female lacks any crimson and is much more streaked. Pairs of birds are often seen undulating across the heath making their familiar *"Tsooeet"* call or a more rapid twitter. The song, which like the Yellowhammer is usually delivered from the top of a bush or overhead wire, is an annoying musical twitter interspersed with sharp whistling or nasal notes. Linnets are common all over the British Isles and occur on most lowland heaths, whilst in winter they are often gregarious and will spend time along coastal marshes forsaking the barren and windswept heath.

A small, neat, grassy nest is built in April and is lined with the down from plants, such as thistles and willows. Four to six eggs are laid in each of two clutches and these are a faintly bluish-white marked with specks and blotches of reddish-purplish-brown, usually clustered at the larger end of the egg.

Yellowhammer *Emberiza citrinella* 17 cm
This pretty little Bunting is quite unmistakeable and can generally be located easily on the heath by its famous song, a rapid "Chi-chi-chi-chi-chi-chweee" which

allegedly says "a-little-bit-of-bread-and-no-cheese". This is delivered from any high point, be it an overhead wire or the top of a gorse bush. Males have a conspicuous lemon-yellow head and underparts and a glowing chestnut rump, their back is streaked with chestnut and their flanks are similarly marked. White on their outer tail feathers also shows conspicuously in flight and juveniles appear like the female which has much less yellow and overall darker markings, particularly on the head. Yellowhammers are common on heathland, roadsides and scrubland countrywide and build a small, grassy nest at the foot of a gorse bush or on the side of a ditch. These are notoriously difficult to find, but if an accidental footfall yields a Yellowhammer's secret you will find the eggs beautifully marked with spots and squiggles of black on a grey-cream base. These markings once gave the birds the colloquial name of Scribe Lark. In winter large parties can be seen scouring the heath, but more often they turn to more profitable areas of foraging such as farmland and even urban scrubland.

Other birds which are often encountered on heathland include: Goldcrests *Regulus regulus* (who are regular winter visitors trying to eke out an existence on the meagre supply of invertebrates hiding in the evergreen gorse) and Whitethroats *Sylvia communis*, which regularly nest on the drier belts of gorse. Lapwings *Vanellus vanellus*, Snipe *Gallinago gallinago*, Redshank *Tringa totanus*, Curlew *Numenius arquata*, Mallard *Anas platyrhynchos* and Teal *Anas crecca* can be found nesting in the wetter valley bogs and pools. A sad loss, in Britain at least, is the Red-backed Shrike *Lanius collurio*, once so common on heathlands, but now decreasing all over its north-western range due to habitat destruction and climatic change. While this has been declining, however, the Reed Bunting *Emberiza schoeniclus* has been expanding its niche and is now becoming a heathland regular.

Common Shrew *Sorex araneus* 11 cm
Of all the small mammals found on British heaths this is by far the most likely to be encountered as the poor selection of plants supports very few herbivores, such as Field or Bank Voles. However, the vast array of insects enjoying the hot climate supports a low density shrew population. Most often you will only hear these insectivorous mammals as they bicker in their territorial disputes, making a loud recurring squeaking from under the heather or grass. If you do catch sight of them they are very dark brown in colour above, but have paler brown flanks which contrast sharply with the grey, yellow tinged underparts. Juveniles are generally lighter above. Their movement is swift and bustling and they manically explore every crevice with their cute, mobile snout and exceptionally sensitive whiskers. Breeding nests are often made under discarded litter on the heath, and if this is overturned in the search for reptiles, shrews can be seen bolting away into the bracken. There are five to seven young in each litter and may be up to five litters in a season, so perhaps not surprisingly these overcharged little animals seldom survive beyond their second autumn.

A GUIDE TO BRITISH HEATHLAND NATURE RESERVES

Bedfordshire

The Firs. TL 028376. 12.4ha.
Bedfordshire County Council.
The best remaining area of heathland left
on the greensands in Bedfordshire. Nothing
super special but smells right in summer
and has all the usual species.

Berkshire

Edgebarrow Woods. SU 837632.
31.2ha. Bracknell District Council.
Has a good range of characteristic species
amongst the plant fauna with Silver-
studded Blue and Grayling butterflies and
Emperor and Fox moths.

Inkpen Common. SU 382641. 10.4ha.
Berkshire, Buckinghamshire and
Oxfordshire Naturalist Trust.
A small common which is now mown to
check scrub encroachment and was in the
past ravaged by fire. Very good for
heathland plants including Pale Dog-
violet – the only site in Berkshire where this
species is found.

Snelsmore Common Country Park.
SU 463711. 58.4ha. Newbury
District Council.
Birds possibly include Nightjars with
visiting Hobbies and Grasshopper War-
blers. A good valley bog with typical flora.

Cheshire

Little Budworth Country Park.
SU 590655. 33ha.
Cheshire County Council.
Typical heathland topography but is
starved of all of the southern specialities.

LEFT **A heathland pool**
RIGHT **A White Water-Lily**

Thurstaston Common. SJ 244853.
75.2ha. National Trust.
Very typical heathland topography with heather, bracken and scrub.

Cornwall

Chapel Porth Nature Trail. SX 697495.
3.2km. Cornwall County Council.
Typical coastal heathland with coastal cliffs and their associated ornithological mayhem.

Isles of Scilly. 1600ha. Duchy of Cornwall/Nature Conservancy Council.
Five inhabited islands, forty more with terrestrial plants have windlashed heaths but the islands are more interesting for their island endemics and many flourishing imports. Don't go in autumn unless you like twitchers.

The Lizard. SW 701140. 400ha.
Various governing bodies.
Coastal cliffs and heathland. This heathland is very special because of its flora. This includes Cornish Heath, Dwarf Rush, Pygmy Rush, Dorset Heath, Chives, Spring Squill, Pale and Common Dog Violet and more typical species such as Bog Asphodel and Sundews. There are areas rich in *Sphagnum* mosses with Pale Butterwort and Lesser Butterfly Orchid. Also Fern Pillwort, Yellow Centuary, Three-lobed Crowfoot and the rare hybrid of Cornish and Cross-leaved Heath *Erica williarnssii*. Altogether a great botanical bonanza.

Devon

Aylesbeare Common. SY 057898.
180ha. Royal Society for the Protection of Birds.
A piece of lowland heath with no access off the footpaths. A satellite site for Dartford Warblers which may be found after prolonged spells of favourable winters.

Nightjar, Curlew and Stonechat are present as is a typical range of heathland flora. This is supplemented by a fine butterfly list of thirty-two species, including Silver-studded Blues and Graylings and a range of dragonflies over the damp bogs.

Chudleigh Knighton Heath.
SX 838776. 74ha. Devon Trust for Nature Conservation.
A lowland heath with a range of topographical features, including some small ponds. Flora includes Common Bladderwort and Pale Dog Violets.

Dartmoor National Park. 94,500ha.
Devon National Park Association.
Some restriction of access because of MOD firing ranges. Varied spreads of heather with typical plants, but also Bristlebent which is only found in the south-west. The bogs hold cotton-grass, Crowberry, Pale Butterwort, Marsh St John's-wort, Ivy-leaved Bellflower and Bog Pimpernel. Many other habitats also occur. Lots of information available.

Dorset

Dorset, with Hampshire, is really the last tenuous sanctuary for many of Britain's 'heathland' species such as Sand Lizards, Smooth Snakes, Dartford Warblers, Nightjars, Hobbies, Marsh Gentians, Ladybird Spiders etc. The major heaths have disintegrated and the shreds are nearly all reserves, many of which have restricted access, but this is true heathland.

Arne. SY 984885. 525ha.
Royal Society for the Protection of Birds.
Permit allows access to main body of the reserve. A marvellous area which holds specialities such as the Small Red Damselfly, Downy Emerald, Scarce and Southern Hawkers in its wetter areas and is famed for having all six of Britain's reptiles and the Dartford Warbler.

Avon Forest Park. SU 128023. 600ha.
Dorset County Council.
Within the guilty coniferous plantations,
spreads of relict heathlands hold Dartford
Warblers, Sand Lizards and Smooth
Snakes.

Cranborne Common. ST 103112.
42.8ha. Dorset Naturalist Trust.
Typical Dorset heath with a permit
required for access to parts off the right of
way. A range of habitats from wet to dry
with a good selection of dragonflies and
birds, e.g. Dartford Warbler and Nightjar.

Hartland Moor*. Restricted access.
258ha. Nature Conservancy Council.
Access is by permit only but this reserve is
a must for the serious investigator. This is
one of the best scraps of Dorset heath
remaining, with all six reptiles and many
rare insects.

Holton Heath*. Restricted access.
80ha. Nature Conservancy Council.
Another very special area where access is
by permit only. The dry heath and its woodland is rich in spiders, wasps and
butterflies (including Graylings and Silver-
studded Blues). Both Sand Lizards and
Smooth Snakes are present.

Morden Bog*. Restricted access.
149ha. Nature Conservancy Council.
Access by permit only. This is dry heath,
bog and overgrown duck decoys isolated
in a sea of coniferous plantations. It is an
exceptional area rich in insect life and
ornithological goodies including Nightjars
and visiting Hobbies.

Studland. SZ 034836. 631ha.
Nature Conservancy Council.
Two nature trails and a reserve centre
(open Sundays or by appointment) provide
access to an aesthetically and ecologically
superb area of Dorset heathland where
Royal Ferns, Great Sundew, Marsh Gentians, Greater Bladderwort and Lesser
Dodder can be found.This reserve also
holds all six British species of reptile, a fine
array of dragonflies and birds such as the
Dartford Warbler and Nightjar. One of the
best areas for public investigation outside
the New Forest.

Tadnoll Meadows. Restricted access.
44ha. Dorset Naturalist Trust.
Primarily of botanical interest with a
species-rich meadowland which includes
Water Avens and Great Burnet. A good
area of *Sphagnum* and Purple Moor-grass
is present and Smooth Snakes can be
found if you are extremely lucky.

Woolsbarrow. SY 892925. 7.6ha.
Dorset Naturalist Trust.
No access away from tracks. An Iron Age
hillfort which is topped by dry heathland, in
a sea of forestry. Noted for its birds
(Nightjar) and insects (Wood Tiger Beetle).

Dyfed

Dowrog Common. SM 769268. 81.2ha.
West Wales Naturalist Trust.
This lowland heath has most of the typical
heathland flora across its range of topography from dry to wet soils. Gorse and Bell
Heather give way to Cross-leaved Heath,
Sphagnum, Bog Asphodel, cotton-grass
and Bogbean, with nearly 300 other plants
also recorded. The Small Red Damselfly is
present as is the Green Hairstreak butterfly. The reserve is ornithologically starved
of the southern specialities, but Merlins
and Hen Harriers visit in winter.

Essex

Tiptree Heath. TL 884149. 24ha.
Tiptree Parish Council.
This little block of heathland seems incredibly out of place in Essex but has all the
typical floral content of this habitat, i.e.
Gorse, Heather, Cross-leaved Heather
with Heath Milkwort and Tormentil. Worth
checking out if you're local.

Greater London

Wimbledon Common. SW19. 341ha.
Gorse and birch scrub can be found with a couple of valley bogs which have Bogbean, Bog Asphodel, sundew and bog mosses.

Hampshire

Broxhead Common. SU 806374. 44ha.
Hampshire County Council.
This patch of dry heathland has mature and regenerating heather atop slopes where gorse and birch encroachment have provided cover for nesting birds. Not glorious but glorious views.

Fleet Pond. SU 816553. 56ha.
Hart District Council.
Half of the reserve is filled with the open water of the pond, which incidentally often attracts vagrant birds, the rest is heathland amongst mixed woodlands. Not a typical 'Hardy' heath.

Ludshott Common Nature Walks.
SU 850360. Ludshott Common
Committee/National Trust.
A range of heathland walks reveal typical heathland plants and a few birds.

New Forest. 37,560ha.
Forestry Commission and others.
Just under half of the 'Forest' is covered by woodland today but between such blocks occur many spreads of grasslands, heathlands and bogs. The range of heathland types is almost unlimited with many ecotypes and their associated communities. Although now only eight per cent is mature heath, rarities are numerous: plants such as Wild Gladioli, Marsh Gentian and Bog Orchids; Hobbies, Woodlarks and Dartford Warbler all nest regularly; and many insects occur, such as the two more Mediterranean damselflies, the Small Red and the Scarce Blue-tailed Damselfly. Less celebrated species also occur: all three species of sundew, Yellow Centaury, Broadleaved and Slender Cotton-grass, Touch-me-not Balsam, Bog Sedge, Slender Sedge, Stonechats, Whinchats, the pipits, a few Wheatears, Nightjars, Tiger Beetles, *Ammophila*, the Sandwasp and Smooth Snakes. Despite its decay the New Forest is one of the most important heathland reserves left in southern England. Access is by road and the many thousands of forestry paths and the scope for exploration virtually endless. This is twentieth-century heathland; don't miss it.

Yateley Common Country Park.
SU 822597. 197ha.
Hampshire County Council.
This north Hampshire common is traversed by a 2km nature trail which has many typical plant and animal heathland species. There is a fine pool which attracts a range of dragonflies and butterflies including the Silver-studded Blue.

Hereford and Worcester

Broadmoor Common. SO 601363.
13.6ha. Hereford and Worcester County Council.
This unusual and untypical splodge of heathland amongst the limestone vales is largely tree and gorse scrub. This reserve has plenty of insects and is renowned for its Spiny Rest-harrow.

Devil's Spittleful. SO 815752. 60ha.
Worcester Nature Conservation Trust.
Perhaps the best piece of 'heathland' in the county, this has typical spreads of heather, bracken and gorse. The grasslands have Harebells and Wild Pansies, but like many 'heaths' away from the south and east of Britain it lacks the panache of these areas. Go south, young man, go south.

Hartlebury Common. SO 820705. 91ha.
Hereford and Worcester County Council.
This lowland heath has bogs and pools

amongst the gorse and heather which have Marsh Cinquefoil and Bogbean. Stonechats and Whinchats are also present.

Hertfordshire

Hertford Heath. TL 350106 and 345111. 25ha. Hertfordshire and Middlesex Trust for Nature Conservation.
Out of the wooded areas a small amount of heather and Purple Moor-grass is spread over the heath which has a few damp areas with pools. Lousewort and Petty Whin occur.

Marshalls Heath. TL 161149. 3.6ha. Hertfordshire and Middlesex Trust for Nature Conservation.
A relict of heathland with typical acid plant life and associated insect fauna.

Kent

Hothfield Common. TQ 972457. 56ha. Kent Trust for Nature Conservation/ Ashford Borough Council.
Now managed to maintain the heathland qualities, this unusual scrap of heath and bog in Kent has many typical heathland species otherwise locally rare, e.g. sundews. Otherwise Ling gives way to Cross-leaved Heath and to cotton-grasses, Marsh Pennywort and Bog Pimpernel. A causeway across the bogs provides access to many of the species and several sand wasps are said to occur.

Leicestershire

Bradgate Park. SK 523116. 320ha. Bradgate Park Trust.
A large area of grassland, heath and bracken with woodlands and rocky outcrops. Leaflets are available from the information centre.

Lincolnshire and South Humberside

Kirby Moor. Restricted access. 56ha. Lincolnshire and South Humberside Trust for Nature Conservation.
Another sore thumb of heathland in an untypical locality, this piece has a wide expanse of heather and Wavy Hair-grass. Blue Fleabane is of interest in the marsh and Bog Pimpernel, Cotton-grass, Skullcap and Marsh Violet occur in an array of sedges and rushes. The wooded pool provides a refuge for dragonflies, and Lesser Redpolls occur on the heath.

Linwood Warren. Restricted access. 26.4ha. Lincolnshire and South Humberside Trust for Nature Conservation.
Access is by permit only to this rare piece of Lincolnshire heath. It has all three common heathers, Round-leaved Sundew, Marsh Violet and Sneezewort. Drier areas have Adders and Nightjars and 200 species of moth and twenty-one butterflies have been recorded.

Scotton Common. Restricted access. 14.5ha. Lincolnshire and South Humberside Trust for Nature Conservation.
An area of both wet and dry heath with Heather, Cross-leaved Heath, Wavy Hair-grass, Bracken and, more interestingly, Marsh Gentian, Heath Spotted Orchid and Royal Fern can be found. Birds include Nightjars and Tree Pipits and both Adders and Common Lizards occur.

Norfolk

East Winch Common. Restricted access. 32ha. Norfolk Naturalist Trust.
A relict of acid heath encapsulated in birch–oakwood, this area has sundews and Bogbean in its mires and has breeding Yellowhammer, Linnet and Nightingales. Adders are common.

East Wretham Heath. TL 914886.
147ha. Norfolk Naturalist Trust.
(For access visitors must report to the
warden at 10 a.m. or 2 p.m. Closed Tues-
days.)
Once modified as a wartime airfield, this
area has two very fine breckland mires.
These are lined with Reed Canary-grass
with Golden Dock and Knotted Pearlwort.
Aquatics include Shining and Fennel
Pondweed, and Amphibious Bistort. The
drier heath is chiefly of Wavy Hair-grass
with heather, Harebells and Heath Bedstraw. A wide range of birds nest but
Nightjar, Whinchat and Wheatear are the
heath's specialities while Hobbies and
Hen Harriers are summer and winter visitors respectively.

Roydon Common. Restricted access.
57ha. Norfolk Naturalist Trust.
Access is by permit only but the reserve
can be overlooked from a public right of
way at TF 680229. This is a typical grada-
tion of dry heath, with its heather, to wet
heath with its Purple Moor-grass, Cross-
leaved Heath, cotton-grass and Deergrass. Adders, Common Lizards and a
variety of insects occur including two
unusual wasps.

Sandringham County Park.
TF 689287. 240ha. Sandringham Estate.
A mixed area of broadleaved and
coniferous woodland and some heathland
which has nesting Nightjars. Nature trails
are provided.

Weeting Heath. Restricted access.
138ha. Norfolk Naturalist Trust/Norfolk
County Council.
A marvellous piece of breckland with
associated plant life: Early Forget-me-not,
Rue-leaved Speedwell and the rarer
speedwells and maiden pink. Observation
hides are available (permit from resident
warden) from April to August to view in
comfort the enthralling Stone Curlews,
which return here annually.

Winterton Dunes. TG 498197. 105ha.
Nature Conservancy Council.
This acid dune system provides one of the
last breeding sites of the Natterjack Toad.
It occurs in the pools which are to be found
in the damp slacks behind the dune ridge.
Here an array of the heathers give way to
Cross-leaved Heath and willows. Ferns are
also of interest: the Broad and Narrow
Bucklerfern grow together with Royal
Ferns. Grey Hair-grass is quartered in
winter by visiting Hen Harriers and the
occasional Rough-legged Buzzard whilst
Little Terns visit the shingle in summer.

Somerset

Exmoor National Park. 68,635ha.
Exmoor National Park Committee.
A mixture of moorland and heathland.
Where it is more 'heathy', heather is domi-
nant with Bracken, Bilberry, Bristle Bent
and Gorse. Wetter sections contain more
Purple Moor-grass, Deergrass, cotton-
grasses and patches of *Sphagnum*. Here
Cross-leaved Heath, Round-leaved Sundew, Bog Asphodel and Pale and Large-
flowered Butterwort are found. Both the
Northern Eggar moth, a heather feeder,
and its southern cogener the Oak Eggar
occur, but the birds are more typically
moorland with the exception of Curlew,
Whinchat and Stonechat.

Langford Heathfield. ST 100236. 72ha.
Somerset Trust for Nature Conservation.
A variety of acid woodland and acidic wet
heath.

Staffordshire

Cannock Chase. SU 971842. 870ha.
Staffordshire County Council.
A relict of the seventeenth-century
heathland, the area has heathers, Bilberry,
Crowberry and an upland species, Cow-
berry. This hybridises with Bilberry and
grows better here than anywhere else in

this country. In the wetter areas Bogbean, Marsh Cinquefoil, Southern Marsh Orchid and Marsh Hawksbeard are found. A few other such moorland/upland species occur, such as Grass of Parnassus, but the accent is primarily southern with the bogs reminiscent of those of Hampshire's New Forest. Locally unusual species such as Bog Asphodel, Bog Pimpernel, Marsh Pennywort, Common Butterwort, Cranberry, Marsh Valerian and Marsh Violet, and the Dioecious and Few-flowered Sedges.

Green Hairstreaks occur with eponymous Emperor, Oak Eggar and beautiful Yellow Underwing moths and these are no doubt hawked by the resident Nightjars and Whinchats. Merlins, Hen Harriers and Great Grey Shrikes are winter visitors. Muntjac, Sika, Red, Roe and the commoner Fallow Deer all graze the heath. This site is really a must for any Midlander with an interest in heathland flora and fauna.

Highgate Common Country Park.
SQ 844900. 111.2ha. Staffordshire County Council.
The expanse of heath is studded with birch woodland and is attractive to a few resident and visiting bird species.

Suffolk

Cavenham Heath. TL 757727. 140ha. Nature Conservancy Council.
Access off the nature trails provided is by permit only. One of the finest remaining breckland heaths with many floral interests. The drier areas of heather have Biting Stonecrop, Thyme-leaved Sandwort, Sand Spurrey, Common Cudweed, Hare's-foot Clover, Mossy Saxifrage and Bearded Fescue. The heather is deep in places and is underlain with lichens and Sheep's Sorrel. Wetter areas have ferns, Common Reed and Meadowsweet, Gipsywort, Water Mint and Lesser Pond Sedge. Linnet and Yellowhammer are common and the rarer Nightjar, Whinchat and

Woodlark also breed. Grayling butterflies and Emperor moths occur.

Dunwich Common. TM 476685.
85.6ha. The National Trust.
This coastal heath or sandling is the most colourful area of heath I have ever seen. All three heathers, Gorse, Western Gorse, Broom and Rosebay Willowherb cause an explosive field of colour between July and September. Sand Sedge and Field Woodrush are also of interest and the views from the car-park over the sea and Minsmere are superb. Indeed, this car-park with its dawn memories of Marsh Harriers and dusk Bitterns booming through the stench of gorse has to be one of my top 'British Bits'.

Knettishall Heath Country Park.
TL 956806. 72ha. Suffolk County Council.
An area of breckland heath and woodland with information on site.

Minsmere. TM 473672. 588ha. Royal Society for the Protection of Birds.
Probably the most famous nature reserve in Britain, it has a range of habitats, including some heathland. For access to this a permit is required. It is very similar to Dunwich Common (see above) with Heather, Bell Heather, Gorse with scrub invasion by birch and pine saplings. In the past Stone Curlew and Red-backed Shrike nested here, but now thankfully at least the Nightjar remains with the commoner Linnet and Yellowhammer.

Thetford Heath. Restricted access.
100ha. Nature Conservancy Council/ Norfolk Naturalist Trust.
There is no access to this magnificent relict of breckland heath between April and July. An array of vegetation types attracts an exciting bird fauna.

Walberswick. TM 493742. 514ha.
Nature Conservancy Council.
A variety of reed beds, mudflats, woodland and some heathland. This has a spread of

Heather Bell, some Cross-leaved Heath and clumps of Gorse and scrub typical of the sandlings (see also Dunwich Common and Minsmere). Nightjars, Stonechats and in winter Great Grey Shrikes occur along with Green Hairstreak butterflies.

Wangford Glebe. Restricted access. 16ha. Suffolk Trust for Nature Conservation.
Access to this reserve is by permit only and is a must for the purist. This is the last patch of the shifting inland dunes which once typified Suffolk's topography; now it's about the size of Trafalgar Square. Sand Sedge laces across the sand and the parched crispy floor of lichen and rabbit holes is a piece of ecological history.

Westleton Heath. Restricted access. 46.8ha. Nature Conservancy Council.
Actual access is by permit only but on-site picnic areas and rights of way allow limited access. Close to Dunwich Common, this too is a fine example of sandling, which has Woodlark and Nightjar as breeding birds.

Surrey

Chobham Common. SU 965648. 198ha. Surrey County Council.
There is a marked invasion of birch and pine on this heath of heather, Purple Moorgrass and Bristle Bent. Wet areas have Hare's-tail, Cotton-grass and Round-leaved Sundew but the real interest is the insects, including two endemics, one species of ant and one spider.

Frensham Country Park. SU 849406. 311ha. Waverley District Council.
A fine example of the Surrey heaths, mainly dry with typical associated fauna and flora, it is suffering devastating Scots Pine encroachment in some areas. A few stray Dartford Warblers and Nightjars occur.

Headley Heath. TQ 204538. 112.8ha. National Trust.
Until the 1930s this was sheep-grazed heathland with a few birches. It was a tank range in the war and this tore up the soil, which provided a superb yet disastrous opportunity for the birch seedlings to become established. Later fire and myxomatosis just about sealed a sad fate for the heath. Now it is managed and the process is in check. Pipits, Linnets, Redpolls and Nightjars occur in the summer and the occasional Great Grey Shrike visits in winter. A good range of butterflies is also recorded.

Horsell Common Nature Walks. TQ 989605, SU 007593, 002605, 015611 each 2.4km. Horsell Common Preservation Society.
The walks traverse sandy heathland which has been severely invaded by birch and Scots Pine. Typical heathland plants and birds can be found.

Lightwater Country Park. SU 921622. 48ha. Surrey Heath Borough Council.
A nature trail winds its way through a range of habitats from woodland to wet acid heathland. Many typical species can be seen from the path.

Thursley. SU 900399. 250ha. Nature Conservancy Council.
The dry heath is of Ling, Bell Heather, Gorse and Dwarf Gorse with broken areas of open sand and birch thickets. White and Brown Sedge, Bog Asphodels and Cranberry in the wetter areas are supplemented by Bogbean, Lesser Bladderwort and Bottle Sedge where there is standing water. Great Grey Shrike winter here and ten raptor species have been recorded, but these are overshadowed by the 10,000 insect species which have been identified. These include Silver-studded Blue and Grayling butterflies and Emperor moths, but of outstanding interest are the twenty-six species of dragonfly, a greater diversity than in any other British site.

Witley Common. SU 936409. 150ha.
National Trust.
A large expanse of typical heathland with
its associated flora and fauna.

Sussex

Ashdown Forest. TQ 432324. 2560ha.
Conservators of Ashdown Forest.
Once a royal hunting forest, this area has a
fine mixture of dry and wet heathland and
its associated invading scrubland valley
bogs. All the typical species can be found
including Nightjars, Hobbies, Grayling,
several fritillaries, Emperor Moths and a
fine selection of dragonflies.

Chailey Common. TQ 386210. 173ha.
East Sussex County Council.
Birch scrub and Bracken have invaded
this area but all three heathers occur with
Petty Whin, Heath Milkwort and Tormentil
and the usual common bird community.

Iping Common. SU 853220. 77ha.
West Sussex County Council.
Cotton-grass laces the bogs which stud
this otherwise dry heath with its inevit-
able birch scrub encroachment. Typical
heather heath grades into woodland at its
southern edge and over 100 spiders have
been identified here, including several
rarities.

Lullington Heath. TQ 545018. 62ha.
Nature Conservancy Council.
A botanically interesting reserve since this
is a piece of chalk heath. Here thin acid soil
overlays the chalk so that neither acid
heath nor chalkland species predominate
– a mix occurs, i.e. Heather may grow
alongside Salad Burnet.

Warwickshire

Sutton Park. SP 103963. 859ha.
City of Birmingham District Council.
Only 10km from Birmingham's city centre,

this reserve suffers enormous public pressure. It is a woodland/heathland mix with
large spreads of acid grassland and
boggy areas lined by Purple Moor-grass,
cotton-grasses and filled with *Sphagnum*.
Plants are a little marsh-orientated with
Marsh Cinquefoil, Marsh Horsetail, Marsh
Marigold and fine strands of Marsh
Orchids occurring. Fox and Emperor
moths and a selection of dragonflies are
also seen over these mires.

Yorkshire

Allerthorp Common. Restricted
access. 6.1ha. Yorkshire Wildlife Trust.
An area of damp heathland with Marsh
Cinquefoil and Marsh Gentian and
Curlew, Whinchat and Nightjar.

Skipworth Common. Restricted
access. 240ha. Yorkshire Wildlife Trust.
A nature trail is provided (which is suitable
for people in wheelchairs) over wet and
dry heath with sundews, Skullcap, Marsh
Bedstraw and Marsh Pennywort.
Nightjars are uncommon yet present and
Marsh Gentians may be found.

Strensall Common. Restricted access.
21.6ha. Yorkshire Wildlife Trust.
One of the few remaining lowland heaths
left in Yorkshire, it has a mixture of dry and
damp heath. Marsh Gentian occurs with
Dog Violet, and Nightjars and most other
commoner heathland birds nest.

* Please note that although access is restricted
to the above three reserves many of the species
that they hold can be found elsewhere in
marginally less precious sites where one is not
faced with the inconvenience of obtaining a
permit.

SOME USEFUL ADDRESSES

British Dragonfly Society
c/o The Secretary
4 Peakland View
Darley Dale
Matlock
Derbyshire DE4 2GF

British Herpetological Society
c/o Zoological Society of London
Regents Park
London NE1 4RY

British Trust for Ornithology
Beech Grove
Tring
Hertfordshire HP23 5NR

Dorset Naturalist Trust
39 Christchurch Road
Bournemouth BH1 3NS

Forestry Commission
(South East and New Forest)
The Queen's House
Lyndhurst
Hampshire SO4 7NH

Lincolnshire and South Humberside Trust for Nature Conservation
The Manor House
Alford
Lincolnshire LN13 3DL

Nature Conservancy Council
Northminster House
Peterborough PE1 1UA

Norfolk Naturalist Trust
72 Cathedral Close
Norwich
Norfolk NR1 4DF

Royal Society for Nature Conservation
The Green
Nettleham
Lincoln LN2 2NR

Royal Society for the Protection of Birds
The Lodge
Sandy
Bedfordshire SG19 2DL

Suffolk Trust for Nature Conservation
Park Cottage
Saxmundham
Suffolk IP17 1DQ

Yorkshire Wildlife Trust
20 Castlegate
York YO1 1RP

FIRE

Fire is today a serious problem which can result in catastrophic damage to our remaining heathlands. In almost all cases it is started by human carelessness, so as more and more people become interested in our heathland flora and fauna and visit our remaining heaths it is important to take every step to ensure their satisfactory survival. Heathland is a naturally parched habitat and as such is exceptionally prone to fire damage through the summer months. Please do not discard any broken glass and be especially careful with your cigarettes and matches if you are a smoker. Check that they are fully extinguished even when you discard them into litter bins, which are sometimes provided. Be vigilant of other people's negligence and report all fires however small to the fire brigade immediately. If fire beaters are readily available do your best to safely control the fire until the brigade arrives. If you love it don't lose it. If you don't love it don't use it.

THE PHOTOGRAPHS

All the photographs in this book were taken using 35mm SLR Canon cameras (A-1 and F-1) in conjunction with the following Canon FD lenses: 28mm, 50mm, 70–210mm zoom, 100mm macro and 500mm F8 reflex. They were all taken using Kodachrome 64 slide film using a tripod and cable release. In some cases filters have been used (primarily 81B, Softner and polarising) to enhance or destroy some aspect of reality.

BIBLIOGRAPHY

Publications marked with an asterisk are specially recommended.

d'Aguilar, J., Dommanget, J. L. and **Préchae, R.** (Ed. Brooks, S.). *A Field Guide to the Dragonflies of Britain, Europe and North Africa.* Collins, London, 1986.
Andrewes, Sir Christopher. *The Lives of Wasps and Bees.* Chatto and Windus, London, 1969.
Appleby, L. G. *British Snakes.* John Baker, London, 1971.
***Arnold, E. N.** and **Burton, J. A.** *A Field Guide to the Reptiles and Amphibians of Britain and Northern Europe.* Collins, London, 1978.

Bristowe, W. S. *The World of Spiders.* Collins New Naturalist, London, 1958.
Brooks, M. and **Knight, C.** *A Complete Guide to British Butterflies.* Jonathan Cape, London, 1982.
Brown, L. *British Birds of Prey.* Collins New Naturalist, London, 1976.

Carter, D. *Butterflies and Moths of Britain and Europe.* Pan, London, 1982.
***Chadwick, L.** *In Search of Heathland.* Dennis Dobson, London, 1982.
***Chinery, M.** *A Field Guide to the Insects of Britain and Northern Europe.* Collins, London, 1972.
***Cramp, S.** and **Simmons, K. E. L.** (Eds.). *Handbook of the Birds of Europe, the Middle East and North Africa.* The Birds of the Western Palearctic (7 volumes – 4 published), Oxford University Press, 1977.

Fabre, J. H. (translated by Stawell, R.). *Fabre's Book of Insects.* Nelson, London.
Fabre, J. H. *Insects.* Paul Elek Nature Classics, 1979.
Foelix, R. F. *Biology of Spiders.* Harvard University Press, London, 1982.
Ford, E. B. *Butterflies.* Collins New Naturalist, London, 1945.
Frazer, D. *Reptiles and Amphibians.* Collins New Naturalist, London, 1983.

***Gibbons, B.** *Dragonflies and Damselflies of Britain and Northern Europe.* Country Life/Hamlyn, London, 1986.
***Goodden, R.** *British Butterflies. A Field Guide.* David and Charles, Newton Abbott, 1978.

Harrison, C. *A Field Guide to the Nests, Eggs and Nestlings of British and European Birds.* Collins, London, 1975.
Heinzel, M., Fitter, R. and **Parslow, J.** *The Birds of Britain and Europe with North Africa and the Middle East.* Collins, London, 1972.
Higgins, L. G. and **Riley, N. D.** *A Field Guide to the Butterflies of Britain and Europe.* Collins, London, 1970.
***Hywel-Davies, J.** and **Thom, V.** (Eds.). *The Macmillan Guide to Britain's Nature Reserves.* Macmillan, London, 1984.

***Jones, D.** *The Country Life Guide to Spiders of Britain and Northern Ireland.* Country Life/Hamlyn, London, 1983.

Lack, P. *The Atlas of Wintering Birds in Britain and Ireland.* T. & A. D. Poyser, Calton, 1986.

*Linssen, E. F. *Beetles of the British Isles* (2 volumes). Warne, London, 1959.

McClintock, D. and Fitter, R. S. R. *A Pocket Guide to Wildflowers.* Collins, London, 1956.

Nature Conservancy Council. *The Ecology and Conservation of Amphibian and Reptile Species Endangered in Britain.* Nature Conservancy Council, London, 1983.

*Peterson, R., Mountfort, G. and Hollom, P. A. D. *A Field Guide to the Birds of Britain and Europe.* Collins, London, 1954.

Phillips, R. *Wild Flowers of Britain.* Pan, London, 1977.

Porter, R. F., Willis, I., Christensen, S. and Nielsen, B. P. *Flight Identification of European Raptors.* T. & A. D. Poyser, Berkhamstead, 1976.

Reade, W. and Hoskins, E. *Nesting Birds, Eggs and Fledglings in Colour.* Blandford, London, 1967.

Sharrock, J. T. R. *The Atlas of Breeding Birds in Britain and Ireland.* T. & A. D. Poyser, Berkhamstead, 1976.

Simms, C. *Lives of British Lizards.* Goose and Sons, Norwich, 1970.

Small, D. (Ed.). *Explore the New Forest: Forestry Commission Guide.* H.M.S.O., London, 1975.

*Smith, M. *The British Amphibians and Reptiles.* Collins New Naturalist, London, 1951.

*South, R. *The Moths of the British Isles* (2 volumes). Warne, London, 1961.

Steward, J. W. *The Snakes of Europe.* David and Charles, Newton Abbott, 1971.

Streeter, D. and Garrard, I. *The Wildflowers of the British Isles.* Macmillan, London, 1983.

*Webb, N. *Heathlands.* Collins New Naturalist, London, 1986.

Whalley, P. *Butterfly Watching.* Severn House Naturalist Library, London, 1980.

Figures in bold type refer to photographs (often with text on same page); figures in italics refer to colour paintings.

RIGHT **Autumn birch**

INDEX